From gene to animal

DATE DUE			

From gene to animal
An introduction to
the molecular biology of
animal development

DAVID DE POMERAI

Lecturer in the Department of Zoology,
University of Nottingham

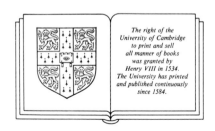

The right of the
University of Cambridge
to print and sell
all manner of books
was granted by
Henry VIII in 1534.
The University has printed
and published continuously
since 1584.

CAMBRIDGE UNIVERSITY PRESS

CAMBRIDGE
LONDON   NEW YORK   NEW ROCHELLE
MELBOURNE   SYDNEY

591.3
D44f
142298
Sept. 1987

Published by the Press Syndicate of the University of Cambridge
The Pitt Building, Trumpington Street, Cambridge CB2 1RP
32 East 57th Street, New York, NY 10022, USA
10 Stamford Road, Oakleigh, Melbourne 3166, Australia

First published 1985
Reprinted 1986

Printed in Great Britain at the University Press, Cambridge

Library of Congress catalogue card number: 84-11332

British Library Cataloguing in Publication Data
De Pomerai, David
  From gene to animal.
  1. Developmental cytology   2. Molecular
  biology
  I. Title
  591.8'761   QH631

ISBN 0 521 26084 1   hard covers
ISBN 0 521 27829 5   paperback

The illustration on the front cover shows compartments in
Drosophila development (see chapter 7 for details). In the larva
(upper part), each imaginal disc (coloured) is subdivided into
anterior and posterior compartments which are destined to form
particular cuticular segments in the adult fly, as shown in the lower
part. Segments are identified as follows: a, clypeolabrum; b,
eye-antenna; c, mandibles; d, maxilla; e, proboscis; I, prothorax
(first legs); II, mesothorax (second legs, wings); III, metathorax
(third legs, halteres); 1–8, abdominal segments; 9–11, genitalia. From
A. Garcia-Bellido, P. A. Lawrence & G. Morata (1979) *Scientific
American* **241**, 90–8. Copyright © 1979 by Scientific American, Inc.
All rights reserved.

PN

In memory of Valerie and Odile,
who would have wished to understand;
and for Lesley,
who will have to try!

Contents

Preface

I have only myself to blame for writing this book. When talking to a publisher's representative, it is unwise to complain that no available text covers all the material one is trying to teach! An invitation from Cambridge University Press to write such a book was forwarded to me while on a study visit to Japan late in 1982, and the basic plan was sketched out during several long railway journeys in that country. In fact, it is based around a series of second- and third-year course units which I have taught for the past few years to students in the School of Biological Sciences at Nottingham University. The whole book was drafted and revised in a period of barely six months (July to December, 1983). I am therefore very grateful to my head of department (Professor P. N. R. Usherwood) and to my research students (especially Andy Carr and Dave Ellis) for their encouragement, and for ensuring a minimum of distractions during that time.

Much of the material dealt with in this book is covered elsewhere in greater detail. Thus chapters 1 to 3 owe a large debt to B. Lewin's *Gene Expression*, vol. 2, *Eucaryotic Chromosomes* (2nd edition, 1980), and chapter 4 to E. H. Davidson's *Gene Activity in Early Development* (2nd edition, 1976). Conversely, a book such as B. Alberts *et al. Molecular Biology of the Cell* (1983) covers a very much wider field. This is not a book about eucaryotic molecular biology as a subject in its own right, but rather about molecular aspects of differentiation and development. By confining the text to animal systems, I have avoided not only plants but also protozoa and yeast (except for very brief mentions); to have covered these topics even sketchily would have demanded a much longer book.

Indeed, my own undergraduate background was in animal development and molecular biology, in the late Professor C. H. Waddington's department of Epigenetics at Edinburgh University. At that time in the early seventies, the subject was a veritable sea of speculation lapping round a few islands of established fact. Thanks largely to the advent of sophisticated techniques for nucleic acid analysis in the later seventies, we now have a kind of submarine with which to explore the sea-bed between those islands. But this is an exploration that has only just begun; only a small fraction of any animal genome (even *Drosophila*) has yet been analysed in any detail, while the rest remains for the moment unread in cloned genomic libraries around the world. Will the next few years bring to light major new surprises, or only further confirmation of what we have already seen?

Neither one nor the other, I suspect; one may doubt whether anything as revolutionary as the discovery of introns is waiting around the corner, but equally the present picture is full of tantalising questions and partial explanations. In some respects this book may be premature, but I believe the future will fill in missing details more often than overturn the basic ideas presented.

This book is aimed mainly at undergraduates taking courses in animal development and/or molecular biology, though I hope others will also find it useful. Those from a traditional zoological background may sense the excitement in molecular biology, discovering how precise gene action underlies the tissues and organisms whose development they study. Conversely, those from a background in genetics or biochemistry may glimpse how genes act together in concert like the instruments of an orchestra, subordinating their individual functions to the needs of the developing organism as a whole. Many indeed are the parallels between musical and animal development, but this is not the place to go into them.

My coverage of the subject matter may seem arbitrary to some. A comprehensive treatise would have necessitated a far longer book, so many molecular details are glossed over. Extensive references to review articles and to recent original papers will, I hope, provide the curious student with a starting point for further reading. The first three chapters deal with topics that may well be familiar to those studying biochemistry or genetics, while chapter 4 repeats material well known to those versed in embryology. To such criticisms I can only reply that my own experience of teaching reveals an appalling ignorance of both molecular and developmental biology among many students in biological disciplines. Both have a bad press in the popular estimation – the one being 'hard and technical' while the other is 'dry and boring'. In trying to cover a large amount of introductory material fairly briefly, I risk superficiality on the one hand and unreadable density on the other. If the resulting compromise is unsatisfactory, that is partly my own fault and partly the inevitable consequence of a rapidly expanding field.

The second part of the book (chapters 5 to 7) may well seem too short.

At one stage, I contemplated several further chapters, for instance on muscle and on cartilage differentiation, on eye development, and on cancer genes. Anything written on the last of these topics would have been out-of-date long before publication, and in any case it falls outside my remit of 'normal' animal development (at least until such time as the functions of endogenous oncogenes have been elucidated in non-cancerous cells). For similar reasons I have not dealt at length with the molecular biology of virus-infected cells. Chapters on muscle and cartilage would have added little that is fundamentally different from the material already covered (chapters 4 to 6), and the same is also true for eye development. But an even more potent reason ruled out this last, since it is my own field of research and hence one which I cannot examine with much pretence of impartiality!

It also follows from this admission that none of the chapters in this book was written with specialist inside knowledge. It was therefore necessary to have them checked over by experts in the fields covered. I am very grateful to all those who helped in this with useful suggestions, constructive criticism and corrections on many points of fact or interpretation. Particular thanks are due to Dr S. R. Barnes (Department of Genetics, Nottingham University; chapter 1), to Dr M. Billett (Department of Biochemistry, Nottingham University; chapter 2), to Dr P. H. W. Butterworth (Department of Biochemistry, University College London; chapter 3), to Dr A. Jeffreys (Department of Genetics, University of Leicester; chapter 5), to Dr N. Carey (Celltech, Slough; chapter 6), to Dr J. R. Tata (National Institute for Medical Research, London; chapter 6), and to Dr P. A. Lawrence (MRC Molecular Biology Laboratory, Cambridge; chapter 7). Warm thanks must go to my teacher, Dr D. E. S. Truman (Department of Genetics, Edinburgh University), who patiently read the entire draft manuscript and responded with much useful advice and encouragement. I have followed virtually all of these reviewers' suggestions, and hope that the text is much improved thereby. Finally, my thanks to the two typists (Mrs Carole Taylor and Mrs Daxa Mehta) who coped with my handwriting and frequent emendations to produce the first and second drafts of the typescript.

David de Pomerai   *Nottingham, September 1984*

Abbreviations and Definitions

rRNA   Ribosomal RNA.

tRNA   Transfer RNA.

mRNA   Messenger RNA.

hnRNA   Heterogeneous nuclear RNA.

rDNA   Ribosomal DNA – i.e. DNA sequences containing rRNA-coding genes.

S value   (For example in 5S, 18S, 28S rRNAs); a non-linear measurement of molecular size based on sedimentation rate.

kb   Kilobases ($=1000$ bases); measurement of RNA chain length.

kbp   Kilobase-pairs ($=1000$ base-pairs or bp); measurement of DNA-fragment or gene length.

MW   Molecular weight.

EM   Electron microscope.

UV   Ultraviolet light.

pol I   RNA polymerase I (A).

pol II   RNA polymerase II (B).

pol III   RNA polymerase III (C).

C-value   Amount of DNA per haploid genome.

Animal pole Region of oocyte containing the nucleus, from which the polar bodies are extruded during meiosis. Also that region of the zygote or early embryo derived from same.

Vegetal pole Region of oocyte, zygote or embryo opposite animal pole; often a yolk-rich part of the egg.

Gene   The classical concept of 'one gene, one polypeptide product' must be reconsidered in the light of recent findings in eucaryotic molecular biology. Certain genes specify stable RNA products (rRNAs and tRNAs) rather than proteins. In some cases, such RNA species remain confined within the nucleus (e.g. snRNAs). To be consistent, one might have to redefine a 'gene' as equivalent to a transcription unit. However, this too produces anomalies. The 18S, 5.8S and 28S rRNAs are separate gene products (i.e. one can speak of the 18S gene as distinct from the 28S gene), yet all three form part of a single precursor transcript. In this case, it is preferable to think of the 18S, 5.8S and 28S units as *cistrons* within a single transcription unit. In the case of protein-coding genes, should the intron sequences (§3.8) be regarded as parts of the gene itself, or as some kind of punctuation between gene segments (exons)? Are the sequences coding for untranslated regions of a messenger RNA chain also parts of the gene? (they are transcribed from exons, but do not code for protein). The discovery of alternative RNA processing pathways (§3.9) means that a single transcription unit can sometimes encode two or more polypeptide products, differing in only parts of their amino-acid sequences. In other instances, a single precursor polypetide can be processed to yield two or more protein units (e.g. lipovitellins and phosvitin from vitellogenin; §6.4).

At present there is no clear consensus as to which parts of a transcription unit constitute the gene proper, and in consequence the term is used rather loosely.

5′,3′   RNA chains are synthesised in a 5′ to 3′ direction. DNA duplexes, consisting of two antiparallel strands, will have both 5′ and 3′ termini at each end of a double-stranded fragment. Conventionally, the 5′ end of a gene is that encoding the 5′ terminus of its RNA transcript. Thus 5′-flanking or *upstream* sequences are those preceding the initiation site for transcription, while 3′-flanking or *downstream* sequences are those which lie beyond the 3′ termination site where transcription is halted. In all cases, 5′ and 3′ refer to the direction in which a gene is transcribed.

PART 1

General

1

Animal DNA

Summary

The genomes of animals contain far more DNA than do those of bacteria (§§1.1, 1.2). Not all of this DNA is informational in the sense of coding for known RNA or protein products. Many DNA sequences are moderately or abundantly repeated within the genome, while others (unique DNA) are present in one or a few copies each (§§1.3, 1.4, 1.5). Highly repetitious DNA is often composed of short simple sequences clustered together in long blocks. These serve no apparent coding function, but their chromosomal locations (e.g. at centromeres) suggest possible structural roles, for example in chromosome pairing. The base composition of such sequences is sometimes sufficiently unusual to permit buoyant-density separation as DNA 'satellites' (§1.4). Other repetitive sequences occur dispersed at many different sites in the genome. DNA reannealing curves yield information on the copy number and sequence complexity of highly repetitious, moderately repetitive and single-copy (unique) classes of DNA in the genome (§§1.3, 1.5). Repetitive and unique sequences of DNA often occur *interspersed* together in a short-period pattern (§1.6), i.e. short repetitive sequences alternating with longer single-copy regions. Long stetches of all-repetitive and all-unique DNA are also found, and this form of organisation predominates in certain insects with small genomes (e.g. *Drosophila*).

Some current techniques for analysing eucaryotic DNA are outlined in section 1.7. These include hybridisation with labelled complementary probes to identify particular sequences, restriction enzyme digestion to cleave DNA strands into defined reproducible fragments, gene cloning to prepare bulk quantities of specific eucaryotic DNA sequences, and two methods for rapid DNA sequencing.

3

Examples of transcribed moderately repetitive sequences include the major ribosomal genes (coding for 18S, 5.8S and 28S rRNAs), the 5S rRNA genes, transfer RNA genes and histone genes (§§1.8, 1.9). Characteristically the ribosomal and 5S genes are each clustered in tandem repeats, with non-transcribed spacer DNA separating adjacent gene sequences (§1.8). Histone genes are also tandemly clustered in lower vertebrates and invertebrates, but in higher vertebrates they occur dispersed as single copies and/or loosely clustered in non-tandem arrays (§1.9). This latter arrangement is also characteristic of many tRNA genes (§1.8).

Like the repetitive histone and tRNA genes, many single-copy protein-coding genes belong to families of related sequences (e.g. globin genes; see section 1.9 and chapter 5). Individual members of such families are often expressed selectively in different tissues and/or at different development stages, i.e. each type of gene within the family encodes a functionally distinct product (e.g. actins; §1.9). Finally in section 1.10, brief mention is made of (i) DNA domains, (ii) transposable (mobile) genetic elements, (iii) DNA amplification and deletion (see also chapter 4), (iv) left-handed Z-form DNA, and (v) the mitochondrial genome.

1.1 Introduction

Living organisms are built up according to inherited instructions encoded in their DNA. It is sobering to remember that this statement – almost a cliché now – would have seemed outrageous to most biologists until a mere 30 years ago. I shall not attempt here to deal with the double-helical structure and replication of the DNA molecule, nor yet with the organisation and control of genes in bacterial and phage systems. These topics, if not already familiar, are covered in detail by many other texts (notably by Watson, 1976, and by Lewin, 1974 and 1977). Rather, this chapter will be concerned with the organisation of DNA sequences in animals, while its two successors will cover the 'packaging' and expression of those sequences.

Descriptions of DNA have frequently resorted to the well-worn analogy of a blueprint or computer program. Though plausible for procaryotic systems, such a comparison could be misleading when applied to higher organisms. It may help to introduce a simple alternative analogy, in terms of a written language where the individual words correspond to *DNA sequence elements* (which may be entire genes, or segments of genes, or non-gene units of DNA). On this basis, a bacterial genome such as the *E. coli* chromosome would read like a concise piece of journalism. No words are wasted and few are even repeated; short sentences or clauses (analogous to operons) are marked off by a minimum of punctuation (initiation and termination sequences). Overall, the information density is consistently high, and comparisons with a computer program are probably appropriate.

But a similar reading of an entire animal genome would be more akin to James Joyce's *Finnegans Wake*. At first sight it seems totally baffling. Words are repeated, fused together, split apart, reassembled in unexpected arrangements; familiar phrases and sentences are interrupted by stretches of apparent nonsense. What proportion of the whole text is truly meaningful remains open to debate, though certainly there are many nuggets of recognisable information (genes) embedded within it.

Much of the rest appears superfluous in the sense that it conveys no useful coding information which we can yet identify. Various speculative functions have been proposed for this 'non-informational' DNA in eucaryotes, but Crick and others have argued that it may be parasitic on the rest of the genome, serving no immediate end other than to perpetuate itself (for review and some assessment of the evolutionary implications, see Doolittle, 1982).

It must be emphasised that eucaryotic DNA is in a state of evolutionary flux – although non-informational sequences generally change far more rapidly than coding sequences (see sections 1.8 and 3.9). This book is concerned specifically with gene expression in animals, and thus tends to treat the genome in a static sense. This is valid insofar as the genome usually remains constant throughout an animal's development (for a few defined exceptions, see section 4.2). Transposable genetic elements (§1.10) can change their chromosomal positions very rapidly, but this does not occur in a predictable fashion consistent with any role in normal development.

1.2 The C-value paradox

Evolutionarily advanced organisms such as vertebrates generally contain more DNA per haploid genome than do simpler animals such as insects or echinoderms, which in turn contain more than unicellular eucaryotes such as yeast or slime moulds (table 1). However, a detailed comparison of eucaryotic genome sizes (known as C-values and expressed in base pairs of DNA per haploid genome) reveals many anomalies. In particular,

Table 1. *C-values of representative organisms (after Lewin, 1980)*

Organism	Group	C-value (bp of DNA)
Escherichia coli	Bacterium	4×10^6
Saccharomyces cerevisiae	Yeast	1.5×10^7
Dictyostelium discoideum	Slime mould	5.4×10^7
Caenorhabditis elegans	Nematode	8.0×10^7
Drosophila melanogaster	Dipteran insect	1.4×10^8
Musca domestica	Dipteran insect	8.6×10^8
Bombyx mori	Lepidopteran insect	5.0×10^8
Aplysia californica	Mollusc	1.7×10^9
Strongylocentrotus purpuratus	Sea urchin	8.6×10^8
Xenopus laevis	Amphibian (toad)	3.1×10^9
Triturus maculatus	Amphibian (newt)	2.2×10^{10}
Necturus maculosus	Amphibian	5.0×10^{10}
Gallus domesticus	Bird (chick)	1.2×10^9
Bovis domesticus	Mammal (cow)	3.1×10^9
Rattus norvegicus	Mammal (rat)	3.0×10^9
Homo sapiens	Mammal (man)	3.3×10^9

the haploid DNA content of many amphibians is much higher than that of mammals, including man (table 1). It seems implausible that more genes would be needed to construct a frog than a human being! At the other extreme, certain Dipteran insects such as *Drosophila* have a much smaller genome than do other members of the same order (e.g. *Musca*; table 1). These anomalies and many others exemplify the C-value paradox.

A different line of argument relating to the C-value paradox is based on the known rates of mutation affecting genes in a variety of organisms. If all of the DNA in higher organisms were informational (primarily protein-coding sequences), then an impossible genetic load would be imposed through the accumulation of lethal mutations. Estimates of the probable numbers of different coding genes required by higher organisms mostly fall within the range of 5000–50 000, which together would occupy far less DNA than that available in most eucaryotic haploid genomes. This implies that a variable proportion of the DNA must be non-coding or otherwise redundant. Particular genes may occur in multiple copies, non-transcribed DNA sequences may play a structural or spacing role in the genome, and there may be a significant fraction of unused or 'parasitic' DNA. Finally, an unknown proportion of the genome will be involved in regulating the expression of other genes.

1.3 DNA abundance classes

The two complementary strands of a double-stranded DNA molecule are held together by A=T and G≡C hydrogen bonding. When fragments of duplex DNA in solution are heated above about 90 °C they denature, so separating the two strands of each fragment. The actual melting temperature (T_m) for DNA varies according to its base composition, since $G{\equiv}C$ base pairs are more stable than $A{=}T$ pairs. On cooling denatured DNA under controlled conditions, hydrogen bonding will occur between complementary sequences so as to reform duplex structures. This is *DNA reannealing* or *renaturation*. For a consideration of renaturation kinetics, the reader is referred to chapter 6 in Davidson (1976). The *rate* of reannealing will depend on a number of factors, most importantly the relative concentrations of the complementary reacting strands. In the case of procaryotic DNAs, where most sequences are represented only once per haploid genome, the observed rate of renaturation will reflect genome size. This gives a sigmoid reannealing curve when the proportion of DNA in duplex form is plotted against the C_{ot} value. This last is a logarithmic function of time, t, and of the initial concentration of single-stranded DNA, C_o (see fig. 1).

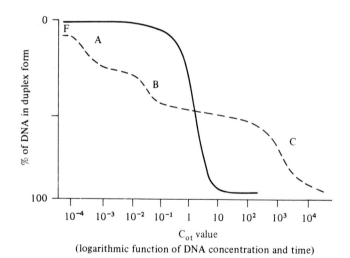

C_{ot} value
(logarithmic function of DNA concentration and time)

——— Bacterial DNA

– – – Typical animal DNA

F   Foldback DNA

A   Rapidly reannealing DNA

B   Intermediate reannealing DNA

C   Slowly reannealing DNA

Fig. 1 DNA reannealing curves (after Britten & Kohne, 1968).

For the DNA of higher organisms with much larger genomes, one might expect a correspondingly slower rate of reannealing, since it will take longer on average for each single-stranded fragment to find a complementary strand from amongst the much wider choice of sequences available. In practice, however, eucaryotic DNAs give a multicomponent reannealing curve (fig. 1; Britten & Kohne, 1968). Some fragments renature rapidly (component A), some at an intermediate rate (component B), and some – as expected – extremely slowly (component C).

The fast (A) and intermediate (B) reannealing components result from the presence of *repetitive* sequences in eucaryotic DNA, i.e. sequence elements occurring in many copies per haploid genome. Single-stranded DNA fragments derived from such repetitive elements will find complementary sequences much more rapidly than will fragments derived from unique elements present only once per haploid genome. A simplified example (fig. 2) will illustrate the principle.

Some means of distinguishing duplex from single-stranded DNA is necessary in order to follow the progress of the reannealing reaction. Three methods are widely used. The first follows changes in the optical density of DNA, which increases during denaturation (hyperchromicity) but decreases due to base stacking during duplex formation. The second method involves chromatography on a matrix of hydroxyapatite (HAP), to which duplex but not single-stranded DNA will bind at moderate concentrations of phosphate ions. The third method uses the S1 nuclease from *Aspergillus*, an enzyme which digests single-stranded but not duplex nucleic acids. Thus the proportion of DNA in duplex form is that which either binds to HAP or is protected from S1 nuclease attack; (a reason for differences between these two estimates is explained below). This information can also be derived from percentage hyperchromicity measurements, relative to 0% for duplex DNA and 100% for single-stranded DNA.

Two further points emerge from figs. 1 and 2. The distinction between components A and B in the eucaryotic C_{ot} curve (fig. 1) presumably reflects differences in the frequency of reiteration between various classes of repetitive sequences. Abundantly repeated sequences (such as x in fig. 2) will find complementary strands more rapidly than will moderately repetitive sequences (such as v in fig. 2). Finally, it will be noted that the eucaryotic reannealing curve does not start at 0% in duplex form. HAP binding suggests that up to 10% of the DNA is already double-stranded at the lowest C_{ot} values attainable, i.e. at zero time (fig. 1). This component is called *foldback DNA* and comprises inverted repeat sequences or *palindromes* which can form hairpin duplex structures by

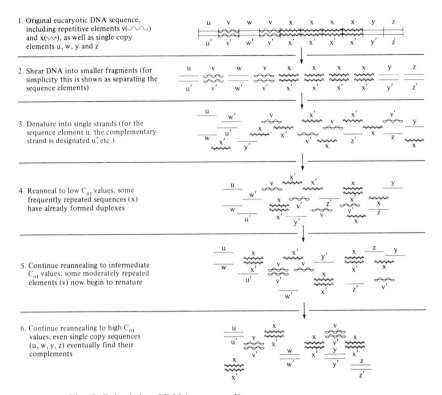

1. Original eucaryotic DNA sequence, including repetitive elements v($\curlyvee$) and x($\wedge\wedge$), as well as single copy elements u, w, y and z

2. Shear DNA into smaller fragments (for simplicity this is shown as separating the sequence elements)

3. Denature into single strands (for the sequence element u, the complementary strand is designated u′, etc.)

4. Reanneal to low C_{ot} values; some frequently repeated sequences (x) have already formed duplexes

5. Continue reannealing to intermediate C_{ot} values; some moderately repeated elements (v) now begin to renature

6. Continue reannealing to high C_{ot} values; even single copy sequences (u, w, y, z) eventually find their complements

Fig. 2 Principle of DNA reannealing.

base-pairing within a single DNA strand (fig. 3). Since the reacting complementary sequences are covalently linked on the same fragment of DNA, duplex formation between them is virtually instantaneous, as it does not depend on random collisions between separate fragments.

Electron microscopy of the DNA renatured at zero time confirms the presence of structures similar to that shown in fig. 3*A*, with duplex stem regions plus single-stranded tails and sometimes loops (Cech & Hearst, 1975). Because these molecules are only partially in duplex form, S1 nuclease digestion will remove the single-stranded loops and tails, giving a lower estimate of the proportion of foldback DNA (1–3%) than does HAP binding. Palindromic sequences of sufficient length to give stable foldback structures are unlikely to occur purely by chance at a frequency representing 1–3% of the total DNA. Among the speculative functions proposed for palindromic sequences is the intriguing suggestion (from A. Gierer) that they may perform a signalling role in the chromosome through conversion into cruciform structures (fig. 3*B*).

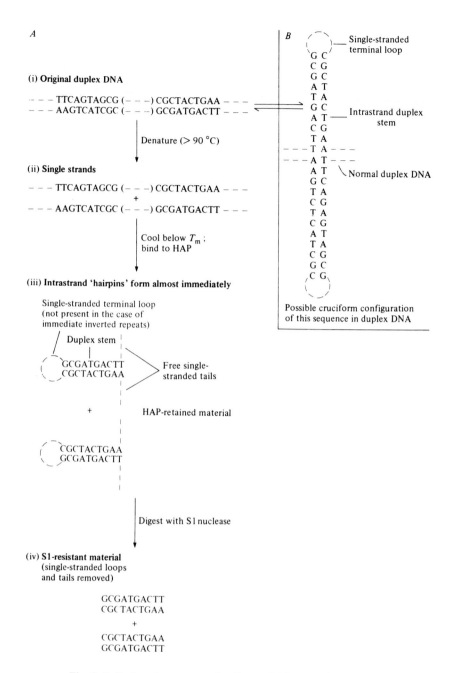

Fig. 3 Palindromic sequence leading to foldback formation.

1.4 Rapidly renaturing DNA sequences

Apart from foldback DNA, a variable proportion of eucaryotic DNA reanneals at low C_{ot} values of less than 10^{-2} (component A in fig. 1). This fraction consists of short simple DNA sequences repeated many times (10^4–10^7 copies) in the genome. Much of the rapidly renaturing DNA is organised into clusters of tandem direct repeats, i.e. adjacent repeat units arranged in a head-to-tail configuration. Two examples from the hermit crab are given in fig. 4.

A

$$\begin{matrix} --- \\ --- \end{matrix}\begin{pmatrix} \text{CCTA} \\ \text{GGAT} \end{pmatrix}_n \begin{matrix} --- \\ --- \end{matrix}$$

B

$$\begin{matrix} --- \\ --- \end{matrix}\left[\begin{matrix} \text{CTGCAT} \\ \text{GACGTA} \end{matrix}\begin{pmatrix} \text{GTC} \\ \text{CAG} \end{pmatrix}_{3-12}\right]_n \begin{matrix} --- \\ --- \end{matrix}$$

Fig. 4 Repeat units of Hermit crab simple sequence DNA (adapted from Bradbury *et al.*, 1981).

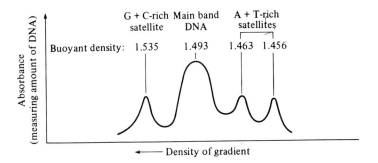

Fig. 5 Separation of satellite and main-band DNAs from guinea pig on Ag^{2+}/Cs_2SO_4 density gradient (after Corneo *et al.*, 1970).

In many cases, the base composition (G≡C:A=T ratio) of simple sequence repeats differs significantly from the average for bulk DNA in that organism. Since G≡C pairs have a higher density than A=T pairs, buoyant-density centrifugation can be used to separate DNA fragments of unusual base composition as 'density satellites' distinct from the bulk of *main band* DNA fragments (fig. 5). Concentrated solutions of caesium chloride are commonly used as the centrifugation medium for buoyant-density determinations, but improved resolution can sometimes be obtained with combinations of caesium sulphate and silver ions (fig. 5). Another method uses the drug Actinomycin D, which binds selectively to GC dinucleotides in duplex DNA, so rendering them less dense.

Most of these satellite DNAs are simple highly repetitious sequences arranged tandemly in long blocks. They are not usually transcribed into RNA (except in lampbrush-stage oocytes; see chapter 4), and rapidly reannealing DNA in general is unlikely to serve any coding function. Satellite DNA sequences are not distributed at random in the genome, but rather occur clustered together in highly condensed (hetero-chromatic) regions of the chromosome. This has been demonstrated by hybridisation *in situ*, as explained briefly below.

Satellite DNAs isolated from insects or mammals are either labelled directly by nick translation (§1.7), or else transcribed *in vitro* using bacterial RNA polymerase and radioactive RNA precursors (§3.2). In both cases a radioactively labelled *probe* (DNA or RNA) results. This is then hybridised with squash preparations of insect polytene chromosomes (chapter 7) or with metaphase spreads of mammalian chromosomes. The nucleic acid reactants are denatured and allowed to interact under conditions which favour hydrogen-bonding between complementary sequences. Excess probe is then washed away and the chromosome preparation covered with photographic emulsion, which will become blackened above those sites where the radioactive probe has bound to complementary DNA sequences on the chromosomes (autoradiog-raphy).

This technique allows one to determine the chromosomal location(s) of satellite or other tandemly repeated DNA sequences. Many mamma-lian satellite sequences are located in condensed heterochromatic regions such as the centromeres of chromosomes (fig. 6). Plausible functions for simple-sequence DNA could therefore include chromosome pairing dur-ing cell division, and/or the maintenance of heterochromatic blocks which inactivate large regions of the chromosome (see Walker, 1971). A role for these sequences in controlling the amount and distribution of recombina-tion has also been proposed (see review by John & Miklos, 1979).

Another feature of rapidly renaturing DNAs is that repeat units within a cluster are not always identical in sequence. In one of the hermit crab satellites (fig. 4B), the number of CAG subunits per repeat can vary from 3 to 12. One consequence of this is that the duplex structures formed by renaturation will tend to be poorly matched, because the reacting strands are not necessarily complementary in sequence along their full length. As a result of this mismatching, the T_m of renatured highly repetitive DNA is markedly lower than that of the original DNA. The rate of mutation among such sequences has been extensively studied in the context of speciation, but this topic is beyond the scope of the present text.

Mammalian DNA also contains highly repetitive sequences dispersed

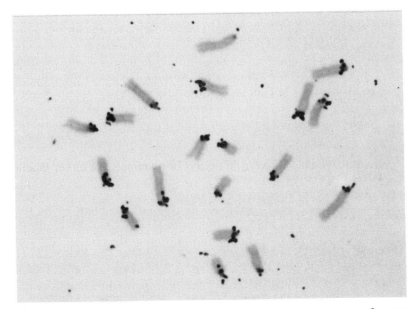

Fig. 6 Location of satellite DNA sequences at the centromeres of mouse chromosomes during metaphase. A radioactive RNA probe complementary to purified satellite DNA was hybridised to a fixed chromosome spread and processed for autoradiography (see text). Black silver grains over the centromeric regions show the location of sequences hybridising with the probe. Magnification ×1620. Photograph by courtesy of Prof. J. G. Gall; reprinted with permission. Copyright 1970, the American Association for the Advancement of Science. From M.-L. Pardue & J. G. Gall (1970) *Science*, **168**, 1356–8.

to numerous different sites within the genome. These are generally recognised as distinct bands when total DNA is digested with particular restriction endonucleases (see section 1.7.2). Short dispersed repetitive sequences less than 500 bp long are known as SINEs (e.g. the *Alu* family in primates), while longer sequences of this type often exceed 5000 bp in length and are known as LINEs (e.g. the *Kpn* family in primates). Typically, SINEs occur in $>10^5$ copies and LINEs in about 10^4 copies per haploid genome. Each family of SINEs or LINEs usually includes several related subfamilies whose sequences and representation in the genome vary widely even between related species (see Singer, 1982).

1.5 Intermediate- and slow-renaturing DNAs

Intermediate-repetitive sequences reanneal at C_{ot} values between 10^{-2} and 10 (component B in fig. 1), i.e. more slowly than the simple-sequence DNA discussed above, but more rapidly than single-copy DNA. They comprise a heterogeneous array of many different types of sequence, ranging in copy number from a few tens to several thousands. Intermediate repetitive sequences also vary considerably as to the lengths of different types of repeat unit; some at least are relatively long, sufficient to include one or more genes within each repeat. Some are known to be transcribed – e.g. the repetitive genes coding for histones and for ribosomal and transfer RNAs – but many are non-coding. Some form tandem clusters with many repeat units arranged in a head-to-tail configuration. Others, however, are more loosely clustered or else occur dispersed at many different sites in the genome. Examples of repetitive genes are discussed in sections 1.8 and 1.9.

Single-copy or unique sequences often comprise the bulk of the DNA in animal genomes. They renature extremely slowly over a C_{ot} range from 10^2 to 10^4 (component C in fig. 1). Essentially these are the DNA sequences present in one or a few copies (<10) per haploid genome. The range of different sequence types represented in this fraction is extremely diverse, but includes the majority of protein-coding genes as well as many non-transcribed sequences. Among the latter are so-called 'pseudogenes', related in sequence to known functional genes, but mutated in various ways so that they cannot be expressed (see section 5.5).

From the renaturation parameters of each component in a eucaryotic C_{ot} curve, as compared to a known kinetic standard such as *E. coli* DNA, one can derive estimates of: (a) the *reiteration frequency* for a typical sequence in that class, i.e. the average number of copies present per haploid genome; and (b) the *complexity* of that class, i.e. the range of different sequences represented therein (conventionally measured in kilobase pairs of DNA). Thus in each haploid genome, the fast-renaturing DNA (component A) comprises vast numbers of copies of low-complexity sequences, the intermediate-renaturing DNA (component B) represents numerous copies of higher-complexity sequences, while the single-copy DNA (component C) consists of one or a few copies each of an enormous range of different sequences. It follows that most of the complexity of a typical eucaryotic genome is contributed by the single-copy (unique) fraction, even though much of the genome bulk may be contributed by repetitive sequences.

For many purposes (especially hybridisation; see section 3.2) it is convenient to deal with single kinetic components of eucaryotic DNA. These may be prepared by stopping the reannealing reaction at different times. Thus fast-renaturing DNA can be isolated as that fraction which binds to HAP at low C_{ot} values (10^{-3} or 10^{-2}), because it is already in duplex form. Unique DNA remains single-stranded at much higher C_{ot}

values (about 10^2), and can then be isolated as that fraction which does not bind to HAP. Several cycles of denaturation and reannealing are used in conjunction with HAP selection in order to prepare purified fractions of simple-sequence, intermediate repetitive or single-copy DNA.

1.6 Interspersion of unique and repetitive sequences in animal DNA

Duplex fragments of uniform length can be prepared from high molecular weight DNA by mechanical shearing, followed if necessary by size-fractionation. The arrangement of repetitive and unique sequences may then be gauged from low C_{ot} reannealing reactions between two preparations of DNA which differ in average fragment length. In order to simplify the experimental system, one of the DNA populations is unlabelled and present in vast excess ('driver' DNA, since its concentration determines the rate of reaction). The other population is radioactively labelled and present in trace amounts (hence termed 'tracer' DNA). After denaturation and reannealing to low C_{ot}, the fraction of input label retained by HAP is measured; this indicates the extent to which tracer strands have formed duplex structures with the excess driver DNA. Usually the driver fragment length is fairly short (e.g. 450 bp) and is kept constant for a whole series of reassociation reactions with labelled tracer preparations differing in average fragment length (e.g. 400, 750, 1500, 3000 bp). Because reannealing is limited to *low C_{ot}* values, only repetitive sequences are able to reform duplexes, while unique sequences remain single-stranded. Any tracer strand containing a region of repetitive sequence will form a partial duplex structure under these conditions, and *all* the label present in that tracer strand should be retained by HAP. Only those tracers composed entirely of unique sequences will remain unbound.

If repetitive sequences are mostly interspersed among unique sequences, one would expect the amount of HAP-bound label to rise with increasing tracer length. On average, repetitive sequences are more likely to occur somewhere within the longer tracer strands than within the shorter ones. The material retained by HAP will nevertheless contain many single-stranded 'tails' of labelled tracer DNA representing the interspersed unique sequences. These tails can be removed by S1 nuclease digestion, resulting in a marked loss of label (fig. 7B). With very short tracer lengths (limiting case), most labelled DNA fragments will consist entirely of repetitive sequences or entirely of unique sequences, but rarely of both together. In this case, the label retained by HAP will be largely S1-resistant (few single-stranded tails; fig. 7A), and will approach

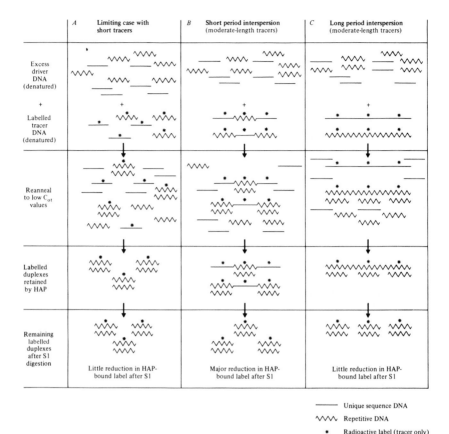

Fig. 7 Principle of determining interspersed sequence organisation.

the proportion of repetitive sequences present in the genome. The same will also be true for longer tracer lengths if the DNA is organised into long blocks of uninterrupted repetitive and unique sequences; again, only a small proportion of tracer fragments will contain both types of sequence together (fig. 7C). Thus as one increases the tracer length, two different results might be anticipated according to the predominant mode of sequence organisation. A marked rise in the amount of HAP-bound label together with an increase in its S1-sensitivity, would indicate interspersion of repetitive and single-copy elements; by contrast, a very slow increase in both parameters would suggest long blocks of all-repetitive and all-unique sequences. These possible outcomes are schematised in fig. 7.

The rate at which the amount of label bound by HAP increases with longer tracer lengths is obviously related to the average size of the single-copy regions which separate neighbouring blocks of repetitive DNA. In the case of *Xenopus* DNA, the following pattern emerges (fig. 8).

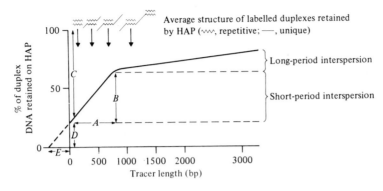

A; average length of interspersed unique sequences.
B; % of DNA that comprises unique sequences with a short-period
   interspersion of repetitive sequences.
C; % of total DNA that is non-repetitive ⎱ intercept values at hypo-
D; % of total DNA that is repetitive   ⎰ thetical 'zero' tracer lengths.
E; average length of interspersed repetitive elements.

Fig. 8 Sequence interspersion in *Xenopus* (modified from Lewin, 1980, after Davidson *et al.*, 1973).

Some 25% of the *Xenopus* genome is repetitive, of which about a quarter (6% of total DNA) forms long blocks of continuous repetitive sequence. These latter are excluded from the interspersion analysis, since they will mostly form duplexes without single-copy tails, whatever the tracer length (fig. 7C). About half of the non-repetitive DNA and most of the remaining repetitive sequences are organised in a *short-period interspersion* pattern, where 300 bp repetitive elements alternate with longer regions of single copy DNA averaging 750 bp in length (Davidson *et al.*, 1973). The parameters of this interspersion pattern can be derived from the intercepts and change of slope in fig. 8; moreover, the average length of the interspersed repetitive elements can be checked by measuring the size of duplex fragments which survive S1 digestion of the material retained by HAP. The remainder of *Xenopus* DNA mostly comprises a longer-period interspersion pattern, where short repetitive sequences alternate with stretches of unique DNA at least 4000 bp long. This is indicated by the much slower increase in HAP-bound label with tracers longer than 750 bp.

The so-called *Xenopus* pattern of short-period interspersion is characteristic of much of the genome in most eucaryotes studied (including molluscs, amphibia and mammals). However, the precise parameters vary somewhat between different animal groups; thus 600 bp repetitive elements interspersed among 2250 bp single-copy sequences characterise some 50% of human DNA (Schmid & Deininger, 1975). By contrast, short-period interspersion is very rare in the genomes of certain insects (e.g. *Drosophila* and the honey bee *Apis*). In *Drosophila* the characteristic pattern is very long-period interspersion; extended blocks of repetitive sequences averaging 5600 bp in length alternate with vast stretches of single-copy DNA at least 13 000 bp long (see Manning *et al.*, 1975). Some interesting evolutionary questions are raised by the observation that not all insect genomes are organised in this way. The DNA of another Dipteran (*Musca*) is organised in the *Xenopus* pattern, with a major component of short-period interspersion. (It will be noted from table 1 that the genome sizes of *Musca* and *Drosophila* are also very different.) Some other organisms such as chick have a genome organisation intermediate between the *Xenopus* and *Drosophila* patterns.

However, the interpretation of these interspersion experiments has been disputed recently on technical grounds (Moyzis *et al.*, 1981a,b). Under stringent reassociation conditions (75 °C), mammalian DNA gives a *Drosophila*-like pattern with very little short-period sequence interspersion, even though the same DNA gives a *Xenopus*-like pattern at 60 °C (Moyzis *et al.*, 1981a). An alternative model in terms of scrambled clustered repeats has been proposed (Moyzis *et al.*, 1981b). Definitive clarification of these interspersion patterns must probably await detailed analyses of long cloned stretches of genomic DNA (see e.g. Shiu-Lee *et al.*, 1977). The dispersed distribution of SINEs and LINEs (§1.4 above; Singer, 1982) in mammalian DNA is also consistent with sequence interspersion.

1.7 Techniques of DNA analysis

Before considering several examples of transcribed repetitive sequences in the eucaryotic genome, it is appropriate to outline some of the major techniques used in DNA analysis (for further details, see e.g. Watson *et al.*, 1983). The preceding sections have already covered DNA reannealing and related methods in sufficient detail.

1.7.1 Probing for specific sequences

The standard method for identifying those fragments of DNA which contain a particular sequence relies on complementary base-pairing between a specific labelled probe and a mixture of unlabelled DNA fragments. The probe may be in the form of RNA, or more conveniently a single-stranded DNA copy of the RNA, known as cDNA (§3.2). In either case a purified preparation of the desired RNA species is prerequisite for generating a specific probe.

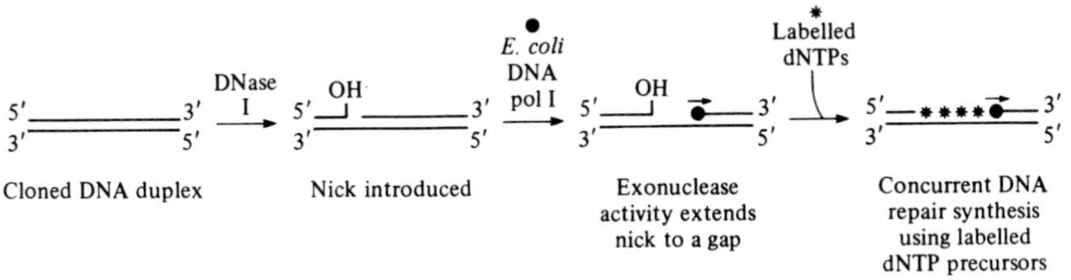

Fig. 9 Nick translation.

Many cloned DNAs (see section 1.7.3) are now available, each representing a specific eucaryotic sequence obtainable pure in virtually unlimited amounts. Double-stranded cloned DNAs are normally labelled directly by 'nick translation' for use as probes. Low levels of DNase I introduce single-stranded breaks (nicks) into both strands of the cloned DNA. *E. coli* DNA polymerase I then extends these nicks into gaps by removing nucleotides from one side of each nick, using its exonuclease activity. The same enzyme simultaneously repairs the gap thus created, by adding appropriate nucleotides onto the other side of the nick, using its DNA polymerase activity. In this way the nick site is moved ('translated') towards the 3' end of the strand. Radioactive DNA precursors (dNTPs) are supplied for the repair synthesis reaction, resulting in highly labelled probe DNA (fig. 9).

Fragments of DNA containing the probe sequence can then be identified by hybridisation with the labelled probe. The most widely used method is to separate the unknown DNA fragments according to size by electrophoresis through an agarose gel. The DNA is then denatured *in situ* on the gel, and the resulting pattern of single-stranded fragments is literally blotted onto a nitrocellulose sheet, which is then baked to immobilise the DNA ('Southern' blotting; Southern, 1975). After hybridisation with the labelled probe, the blot is washed thoroughly to remove excess probe and then autoradiographed. Any DNA fragments containing part or all of the probe sequence will form duplex structures with complementary strands of the labelled probe, and silver grains blackening the final autoradiograph will mark the position of bands containing such fragments. The problem of generating defined fragments of eucaryotic DNA has been solved by the use of restriction enzymes, as dealt with in the following section.

1.7.2 Restriction analysis of DNA

A type II restriction endonuclease represents one half of a bacterial modification-restriction system designed to degrade foreign DNA (e.g.

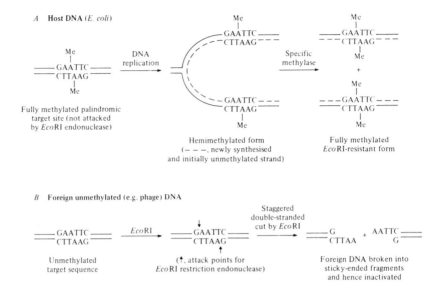

Fig. 10 *Eco*RI restriction system.

from an invading phage) while protecting its own genome. The other half of the system is a sequence-specific methylase which methylates one particular base on both strands within a short palindromic sequence element. The restriction endonuclease recognizes this same palindromic element, but can only effect a double-stranded cut within that sequence when the methyl groups are absent, as is likely to be the case for invading foreign DNA. The example of the *Eco*RI restriction enzyme from *E. coli* will serve to illustrate the principle (fig. 10).

This palindromic sequence is sufficiently short for it to occur about once every 3–4000 base pairs in DNA of average base composition. Restriction enzymes from other bacterial species recognise different target sequences involving 4 to 8 base-pair palindromic elements, and may thus cut DNA more or less frequently than *Eco*RI. The cuts made on both strands are often staggered, generating 'sticky ends' (fig. 10). Other enzymes cut in the centre of the palindrome, giving flush ends. A few restriction systems recognise non-palindromic sequences or separated palindromic elements (see Appendix A in Watson *et al.*, 1983).

What is true for invading phage DNA is also true for eucaryotic genomes. On average *Eco*RI will cut once every 3–4000 base pairs within the huge DNA strands that make up chromosomes. Because the DNA complement is by and large identical in all cells of an individual, it follows that a mass DNA preparation will be cleaved reproducibly by any given

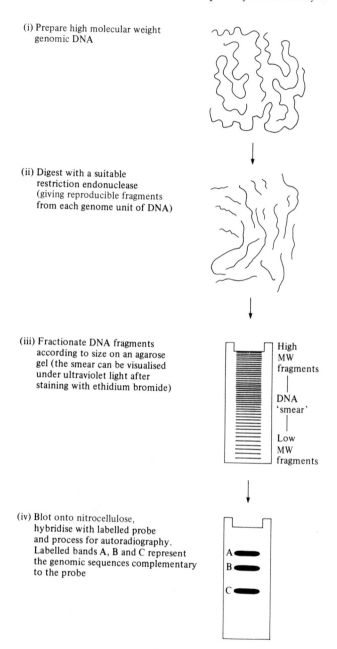

(i) Prepare high molecular weight
   genomic DNA

(ii) Digest with a suitable
    restriction endonuclease
    (giving reproducible fragments
    from each genome unit of DNA)

(iii) Fractionate DNA fragments
     according to size on an agarose
     gel (the smear can be visualised
     under ultraviolet light after
     staining with ethidium bromide)

High
MW
fragments

DNA
'smear'

Low
MW
fragments

(iv) Blot onto nitrocellulose,
    hybridise with labelled probe
    and process for autoradiography.
    Labelled bands A, B and C represent
    the genomic sequences complementary
    to the probe

A

B

C

Fig. 11 Schema for identifying genomic DNA fragments containing
specific sequences.

restriction enzyme into fragments whose lengths represent the distances between recognition sites for that enzyme. This in itself may not be much help, since the size of eucaryotic genomes is so great (often $>10^9$ bp; see table 1) that perhaps a million different fragments will be generated, well beyond the resolving power of any separation system yet devised. Nevertheless, certain restriction enzymes generate distinct bands over and above the general smear of different-sized DNA fragments. These bands result from cleavage sites located within repetitive-sequence elements occurring in numerous copies per genome, whether tandemly clustered or dispersed to many different locations. Examples of the latter include the *Alu* family of SINEs and *Kpn* family of LINEs in primate DNA (see section 1.4). However, the use of radioactively labelled probes (§1.7.1) enables one to identify those restriction fragments containing any unique sequence for which a suitable probe is available. The restriction fragments from genomic DNA are simply separated according to size on an agarose gel, followed by blotting, hybridisation with labelled probe and autoradiography (see fig. 11). This is the basis of restriction mapping, and can also be used on a preparative scale to isolate a size class of DNA fragments enriched in one particular sequence, so providing a starting point for gene cloning; both are discussed in the next section.

1.7.3 Gene cloning

The bands of genomic DNA which hybridise with a given probe after restriction and gel analysis (fig. 11) will be far from pure, owing to the presence of many other sequences which happen to be cut into similar-sized fragments. Nevertheless, a preparative-scale version of the protocol shown in fig. 11 can effect a several-hundredfold purification of particular sequences. Either this partially purified genomic DNA, or else double-stranded cDNA copies of a known RNA sequence (see section 3.2) may be used as the starting material for cloning.

The ultimate aim in this is to introduce eucaryotic DNA sequences into a bacteriophage or plasmid 'vector' which can then be propagated in a suitable bacterial host. Many bacteria contain one or more types of *plasmid* in the form of small extrachromosomal DNA circles which have the following characteristics: (a) they commonly carry drug-resistance markers (thus the widely-used pBR 322 plasmid contains two genes conferring resistance against ampicillin and tetracycline respectively); (b) they can be transferred from one bacterial cell to another (transformation); and (c) they can often replicate independently of the host cell chromosome (for instance, some plasmids continue to replicate when chromosomal replication is inhibited by chloramphenicol, resulting in numerous plasmid copies per cell). Many ingenious methods have been devised for introducing eucaryotic sequences into phage or plasmid vectors. In brief, these mostly involve the production of complementary sticky ends on the vector and eucaryotic DNA components. This necessitates first of all breaking open the plasmid DNA circle with a single restriction enzyme cut. For pBR 322, a *Hind*III site in the tetracycline-resistance gene or a *Pst*I site in the ampicillin-resistance gene are ideal.

Once the plasmid has been linearised, the enzyme terminal deoxynucleotide transferase is used together with a single type of DNA precursor (say dATP) to generate single-stranded *tails* of poly(dA) extending from both 3′OH termini of the plasmid. The partially-purified eucaryotic DNA fragments are treated in the same way, but using the complementary DNA precursor (in this case dTTP) to produce poly(dT) tails. Upon admixture of the two tailed components, a considerable proportion should reanneal to form circular structures held together by A=T base-pairing between the tails (fig. 12). These circles will still contain single-stranded gaps and nicks; the gaps can be filled in using *E. coli* DNA polymerase I (as in nick translation), while the nicks can be sealed by DNA ligase. This will give recombinant closed-circular plasmids containing a eucaryotic DNA sequence inserted within one of the drug-resistance markers of the original plasmid.

The resultant recombinant plasmid shown in fig. 12 should confer ampicillin-resistance but not tetracycline-resistance on any host bacterium (because the eucaryotic DNA now splits the *tet*^{r} gene). It is thereby distinguished from any other circular molecules which may form during annealing; for instance, unchanged (non-recombinant) plasmids will confer both types of drug resistance (*tet*^{r} gene uninterrupted). The preparation of sealed circular molecules is introduced at low multiplicity (<1 plasmid per cell) into a population of bacteria sensitive to both drugs, which are then plated out at low density to give *clonal* colonies, each founded by a single cell. These colonies are replica-plated onto media containing ampicillin or tetracycline or both, so as to identify tet-sensitive but amp-resistant colonies likely to contain recombinant plasmids. The appropriate colonies are then picked off the original plate and grown up on a fresh plate in a grid pattern for ease of relocation. By touching a nitrocellulose filter onto the surface of this dish, a proportion of the cells from each clonal colony will be transferred to the filter in the original grid pattern (as in replica plating). These cells are lysed and the released DNA immobilised on the filter for hybridisation with a suitable labelled probe. Autoradiography will then identify those clones which contain a recombinant plasmid carrying the desired eucaryotic DNA sequence, as opposed to any other contaminating sequence. It is then a simple matter

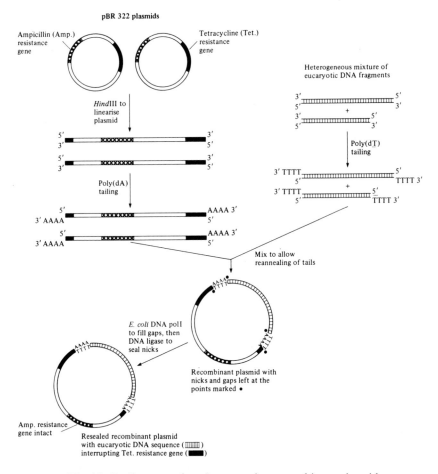

Fig. 12 Outline procedure for preparing recombinant plasmids.

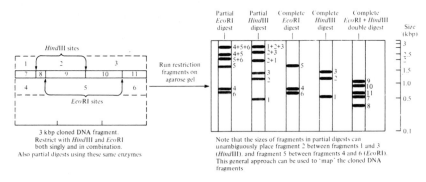

Fig. 13 Restriction enzyme analysis.

to grow up large quantities of bacteria from the positive clone(s), incubate them in chloramphenicol medium so as to increase the plasmid yield, and finally isolate the recombinant plasmid DNA with its cloned eucaryotic sequence. This cloned DNA can then be analysed in detail by *restriction mapping*. The use of two or more restriction enzymes alone and in combination to generate overlapping patterns of different-sized DNA fragments is best illustrated by a hypothetical example (fig. 13). Note the use made of partial digests which cleave the DNA at only some of the available restriction sites.

Analysis with further restriction enzymes in the same way allows the various fragments to be assigned an unambiguous map order within the cloned segment of DNA. However, much more detailed information about a cloned DNA segment can be derived by DNA sequencing (next section).

1.7.4 DNA sequencing

Of the several rapid sequencing methods now available, one of the most widely used is that of Maxam & Gilbert (1977), since it can be applied

A **Preparation of duplex DNA fragments labelled at one 5′ terminus**

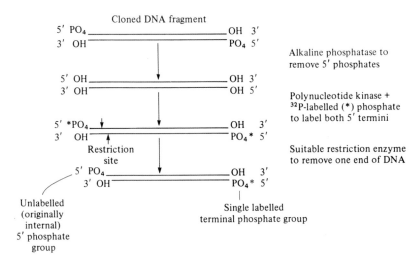

The sequence determined in part *B* will be that of the strand terminating at its 5′ end in the labelled phosphate group. All fragments derived from the other strand will be unlabelled and hence not detected on the final autoradiograph

Fig. 14 Sequencing DNA.

Fig. 14 (contd).

B Partial cleavage method (Maxam & Gilbert, 1977), modified after Bradbury *et al.* (1981)

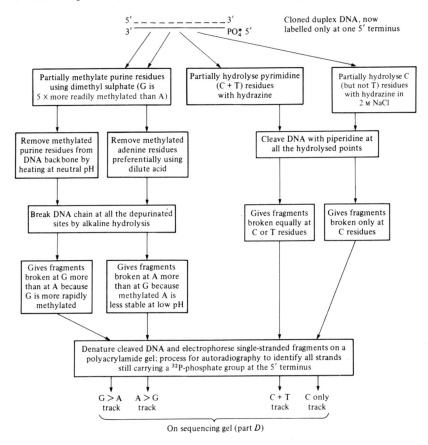

On sequencing gel (part *D*)

directly to double-stranded DNA. Suitably-sized pieces of cloned DNA are first treated with alkaline phosphatase to remove the 5' terminal phosphate groups; these are then replaced with (^{32}P)-labelled phosphate groups using the enzyme polynucleotide kinase. It is then necessary to remove one of the two (^{32}P)-labelled termini from each molecule, usually by means of a restriction enzyme known to cut at a single site close to one end (fig. 14*A*). The DNA molecules – now labelled only at one terminus – are subjected to *partial* cleavage by four methods which break DNA strands respectively at A, G, C or T bases (with greater or lesser specificity; details in fig. 14*B* above). Each method will generate a series of fragments of different sizes because cleavage occurs at only some of the

Fig. 14 (contd).

C   **Specifically terminated labelled fragments generated by partial cleavage
(Maxam & Gilbert, 1977) or dideoxy synthesis (Sanger *et al.*, 1977)**

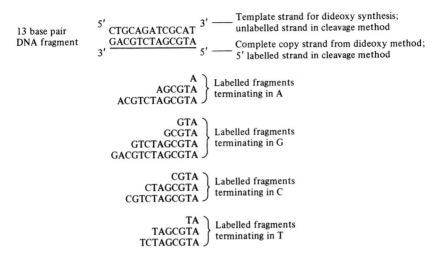

Note that partial cleavage fragments from the Maxam & Gilbert
method will only be labelled if they carry the 5' ^{32}P-labelled
terminus (A). In the Sanger *et al.* method all copy chains will be
initiated with a 5' A residue and will be labelled throughout, but
will terminate at their 3' ends in specific dideoxy-nucleotide
residues

D   **Sequencing gels** (for the 13 base sequence underlined in part *C*).
Each series of specifically terminated fragments (*C*) is size-separated on a gel
system able to detect one-nucleotide differences in chain length

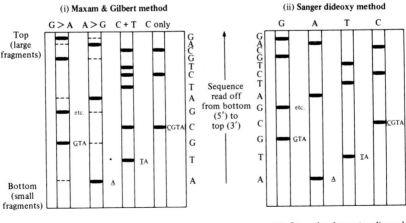

available sites (fig. 14*C*). Those fragments terminating at one end in the (^{32}P)-label and at the other in a specific residue (A, G, C or T) can be identified by autoradiography, after first denaturing the DNA and then separating the single-stranded fragments according to size on a long polyacrylamide gel. Sequences of up to several hundred bases can be read off directly from the autoradiograph (starting with the shortest fragments at the bottom and working upwards), by comparing the presence or intensity of bands generated by the four cleavage methods (fig. 14*D*). In practice, the length of sequence which can be read off is limited by the resolving power of the gel for larger fragments, which become closely spaced towards the top.

An alternative technique devised by Sanger *et al.*(1977) generates a similar series of labelled fragments by partial synthesis rather than degradation. This method necessitates a single-stranded DNA template, which can be provided from sequences cloned in a vector such as the single-stranded DNA phage M13. A primer must also be annealed to the 3′ end of this template, in order for *E. coli* DNA polymerase I to initiate synthesis of a complementary DNA strand (starting from the 3′OH terminus of the primer; cf. gap-filling during nick translation, fig. 9). In the presence of radioactive DNA precursors, the copy strands will be highly labelled, and can thus be detected by autoradiography after gel electrophoresis. In order to use this approach for sequencing, it is necessary to halt the synthesis reaction at specific residues; this can be done by including a proportion of one dideoxynucleoside triphosphate (e.g. dideoxy-ATP) in each of four reaction mixtures. Because each dideoxy-NTP lacks a 3′OH group, it is a chain-terminating residue, i.e. no other dNTP can be joined onto it. For instance, with a mixture of dideoxy-ATP and normal dATP (plus dCTP, dTTP and dGTP), the synthesis reaction will terminate at discrete points defining each of the A residues in the copy strand. Sometimes the polymerase will use dideoxy-ATP at the first site where an A residue is required, resulting in premature termination at that point. But more often a normal dATP will be used at that site, allowing synthesis to continue to the second A residue (or third, or fourth etc.) – i.e. until dideoxy-ATP terminates the chain. Separate reactions using dideoxy-GTP, dideoxy-CTP or dideoxy-TTP will each provide a similar series of *partial* copy strands, terminating in G, C or T residues respectively. Each mixture of specifically-terminated labelled fragments is then separated according to size on one track of a sequencing gel, allowing the sequence to be read off directly from bottom to top on the autoradiograph (fig. 14*C* and *D*).

1.8 Ribosomal genes and tRNA genes

Having summarised some current techniques used for the analysis of DNA sequences, we can now consider specific eucaryotic genes in greater detail, beginning with examples of repetitive genes (§§1.8 and 1.9); single-copy genes are considered separately in chapter 3.

Both ribosomal subunits have an RNA framework provided by ribosomal RNAs (rRNAs). In each small 40S subunit there is a single species known as 18S rRNA on the basis of its sedimentation coefficient. The large 60S subunit includes two small rRNA species designated 5S and 5.8S as well as the much larger 28S rRNA. The approximate sizes of these rRNA species in nucleotides are 120 (5S), 160 (5.8S), 1700–2000 (18S) and 4000–5000 (28S) for most animal species. In sea urchins and some other invertebrates the largest rRNA species is rather smaller, i.e. about 26S (3200 nucleotides). The genes coding for all these rRNAs are tandemly repeated, but fall into two distinct classes, namely the 18S + 5.8S + 28S genes, and the 5S genes.

1.8.1 18S, 5.8S and 28S rRNA genes

In higher eucaryotes these three genes form parts of a single large repeat unit, spanning between 10 and 40 kilobase pairs (kbp) of DNA in different animal groups (higher in mammals). The chromosomal DNA segment carrying the tandem repeats of this unit is defined cytologically as the nucleolar organiser, since it runs through the core of the nucleolus (which is the intranuclear organelle responsible for rRNA production and ribosome assembly). The organisation of a typical rDNA repeat unit, comprising both transcribed and non-transcribed sequences, is shown in fig. 15; details of rRNA transcription and processing will be given in chapter 3.

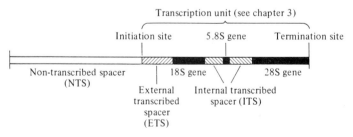

Fig. 15 rDNA repeat unit.

The number of chromosomal copies of the rDNA repeat unit varies from about 40 per haploid genome in the insect *Chironomus* to about 450 in the toad *Xenopus*. Tandem arrays of rDNA units are organised into clusters, sometimes at a single site per chromosome set (e.g. in *Xenopus*), but sometimes at several such sites (e.g. five in humans). Although the transcribed portion of the rDNA repeat varies somewhat in length and sequence between different species, far greater variations occur in the non-transcribed spacer. In the closely related *Xenopus* species *X. laevis* and *X. borealis* (Brown *et al.*, 1972), the rRNA-coding regions are almost identical, whereas the rDNA spacers are quite disparate in sequence.

In the young oocytes of many species, extra copies of the ribosomal genes are produced by a process of selective gene amplification. In *Xenopus* this results in several thousand DNA circles, each containing many rDNA repeat units. These in turn give rise to the extrachromosomal nucleoli which facilitate synthesis of vast amounts of rRNA during oogenesis (§§4.2 and 4.4).

Mutants are known in which the tandem arrays of rDNA units are partially deleted (*bobbed* mutants in *Drosophila*) or even largely absent (o_{nu} homozygotes in *Xenopus*). In the latter case, heterozygotes ($o_{nu}/+$) are viable even though they carry only half the normal number of rDNA units, but in homozygous offspring (o_{nu}/o_{nu}) this mutation is lethal at the tadpole stage (§4.5). *Bobbed* mutants containing more than about one-third of the wild-type rDNA complement are apparently normal, while those with less than this amount show increasingly mutant phenotypes.

1.8.2 5S rRNA genes

In bacteria and lower eucaryotes such as yeast, one copy of the 5S gene is located within each repeat of the rDNA unit. In higher eucaryotes, however, the 5S genes occur tandemly in clusters located at chromosomal sites distant from the nucleolar organiser(s) containing the other rRNA genes. The RNA polymerase which transcribes the 5S genes (pol III) is different from that which transcribes the other rRNA genes (pol I; see section 3.3). Presumably some form of quantitative regulation coordinates the rate of RNA production from these two sets of genes, since equimolar amounts of 5S, 5.8S, 18S and 28S rRNAs are required in order to assemble functional ribosomes.

5S genes are organised in clusters of tandemly repeating units which range in unit length from 400 to 750 base pairs in different animals. Each repeat unit consists of non-transcribed spacer plus a transcription unit only slightly longer than the final 5S rRNA product. Thus relatively little external transcribed spacer and no internal transcribed spacer occurs within the 5S transcription unit (compare the rDNA repeat unit in fig. 15). In *Xenopus laevis* there are two main classes of 5S genes, differing in only 6 out of 120 base pairs of coding sequence. The *oocyte-type* 5S genes are present in about 24 000 copies (showing some heterogeneity) and are expressed exclusively in oocytes. By contrast, the *somatic-type* 5S genes are present in far fewer copies (about 400); these alone are expressed in all somatic cells, where the oocyte-type 5S genes are inactive. A similar distinction applies in *X. borealis*, though in this species the total number of 5S genes is only about 9000. The fact that both species reserve the majority of their 5S genes for use during oogenesis, may represent an alternative to the process of rDNA amplification (see section 1.8.1 above) in order to supply the huge numbers of ribosomes found in the mature oocyte. Other organisms generally contain smaller numbers of 5S genes; less than 200 per haploid genome in *Drosophila* and about 2000 in man. As to their location, 5S gene clusters are found close to the telomeres of all 18 chromosomes in *X. laevis*; by contrast, only a single cluster is present in *Drosophila* (on chromosome 2) and in higher primates including man (on one arm of chromosome 1).

In the oocyte-type 5S repeat units of *X. laevis* each non-transcribed spacer includes a G≡C rich region, 100 bp of which are closely related in sequence to the actual 5S gene. However, this sequence is not transcribed *in vivo* and is therefore classed as a pseudogene (Jacq *et al.*, 1977; see fig. 16). The remainder of the spacer is mostly A=T rich, comprising numerous varied versions of 15 and 24–25 bp subrepeats (Fedoroff & Brown, 1978); apparently the number of subrepeats can vary even between adjacent 5S repeat units, resulting in heterogeneous spacer lengths (Carroll & Brown, 1976). Once again, the 5S-coding sequences are much more closely conserved in evolution than are the spacer sequences separating them. In *X. borealis*, the 5S-coding sequences are very similar

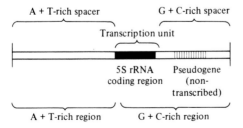

Fig. 16 5S repeat unit (*Xenopus laevis* oocyte type; after Jacq *et al.*, 1977).

to those of *X. laevis*, but the spacers are unrelated. At least some of the *X. borealis* oocyte-type 5S genes are located only 80 bp apart, forming subclusters separated from each other by longer spacers (Korn & Brown, 1978).

1.8.3 Genes coding for tRNAs and other small RNAs

The fact that tRNA genes are transcribed by RNA polymerase III (see chapter 3) relates them to the 5S genes discussed above, though their mode of organisation is rather different. Most animals have between 800 (*Drosophila*) and 6000 (*Xenopus*) copies of the tRNA coding sequences. However, this figure includes genes coding for all of the different tRNA species required to recognise sense codons in mRNA. In general there are between 10 and 200 copies of each type of tRNA gene, reflecting in part the extent of codon use. At one time it was thought that the genes encoding each tRNA species might occur tandemly in small isogenic clusters, but this now appears not to be the case. Detailed analysis of cloned DNAs containing tRNA-coding sequences often reveals several different tRNA genes linked within the same DNA fragment.

In the case of *Drosophila* tRNA gene cluster spanning 9.3 kbp of DNA (Hovemann *et al.*, 1980), there are sequences coding for one $tRNA_2^{Arg}$, three $tRNA^{Asn}$, one $tRNA^{Ileu}$ and three $tRNA_2^{Lys}$. These genes are not all transcribed in the same direction and they are irregularly spaced (no simple repeating spacer). Possibly tRNA genes might occur in these loose non-tandem clusters for convenience in transcription, but single copies of these genes are also found dispersed at other sites (as for instance in yeast). The occurrence of short intervening sequences in some tRNA genes will be covered in section 3.6.

Before leaving the subject of genes coding for small stable RNAs, some mention should be made of those which encode the small nuclear (sn) RNAs. These are abundant stable RNA molecules, designated U1 to U6; they range in size from 107 to 214 nucleotides, and are confined to the nucleus (reviewed in Busch *et al.*, 1982). Some of the snRNAs (particularly U1) may play a key role in RNA processing, a topic to be considered in chapter 3 (see section 3.9).

The snRNA genes of chicken (U1) and man (U1, U2 and U3) are not all clustered; numerous copies of these genes also occur at widely dispersed sites in the genome. Most of the snRNA genes analysed to date are mutated in various ways and are not expressed, implying that there are many more inactive pseudogenes than active snRNA genes (see Roop *et al.*, 1981, and Denison *et al.*, 1981). Three human pseudogenes related to the U1, U2 and U3 snRNAs respectively, are flanked by direct repeats of 16–19 base pairs (van Arsdell *et al.*, 1981). It has been suggested that such dispersed pseudogenes might have arisen through reintegration into the genome of duplex DNA copies derived from snRNA transcripts (see also §5.5). Although the snRNAs are similar in size to the 5S rRNA and 4S tRNAs, they are in fact transcribed by RNA polymerase II rather than polymerase III.

1.9 Repetitive mRNA-coding genes

1.9.1 Histone genes

The five major histones (designated H1, H2A, H2B, H3 and H4) are small basic proteins which associate tightly with DNA in the chromatin of eucaryotic nuclei. The structure and organisation of chromatin will be dealt with in the next chapter. At this juncture it is sufficient to point out that large amounts of all five histones will be required to associate with the newly replicated DNA during cell division, a problem which is particularly acute during the rapid divisions of early cleavage (§4.5). All five major histone genes are repetitive in eucaryotes, with copy numbers ranging from several hundred of each type in sea urchins, to 40 or so in *Xenopus* and mammals, and perhaps as few as 10 in chicken. Histone genes have been cloned from several sea urchin species; they are characteristically organised into clusters, with a large 6–7 kbp repeat unit containing one copy of each of the five major histone genes. The organisation of histone-coding sequences and spacers within the *Lytechinus pictus* repeat unit is shown in fig. 17 (Cohn & Kedes, 1979), together with less detailed information for the *Drosophila* repeat unit (Lifton *et al.*, 1977).

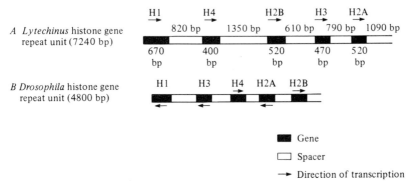

Fig. 17 Histone gene repeating units (after Lewin, 1980).

The order and polarity of the *L. pictus* histone genes within this repeat unit is the same as that in other sea urchins examined (e.g. *Psammechinus miliaris* and *Strongylocentrotus purpuratus*); notably, the five genes are all transcribed from the same DNA strand. In the repeating unit of the histone gene cluster from *Drosophila*, by contrast, the order of the five histone genes is altered, and two of them are reversed in polarity with respect to the other three (i.e. transcription occurs from opposite DNA strands in different parts of each repeat).

In sea urchins a number of histone variants are known, particularly for H1, but also for H2A and H2B. These variants are expressed sequentially during development, and they are encoded by distinct mRNAs transcribed from different genes. There is recent evidence (Kedes & Maxson, 1981) that those histone genes expressed early in development occur at high copy number in tandem repeats (as above; fig. 17), while the genes encoding later-expressed variants occur in fewer copies at scattered locations.

In higher vertebrates (including chick and man; Engel & Dodgson, 1981; Heintz *et al.*, 1981), this tandem arrangement of repeats containing all five histone genes breaks down. Those vertebrate histone genes which do occur in clusters are not organised in tandem repeats, while other histone genes are dispersed as solitary copies separated by 10 kbp or more of non-histone DNA. An intermediate form of organisation is found in the newt *Notophthalmus* (Stephenson *et al.*, 1981); here, all five histone genes are located within a 9 kbp repeat unit present in some 800 copies per haploid genome. The main difference from the sea urchin situation is that each repeat unit is separated from the next by at least 50 kbp of repetitive satellite DNA. Histone genes, like all other mRNA-coding genes, are transcribed by RNA polymerase II (see chapter 3).

1.9.2 Other gene families

The tRNA and histone genes discussed above are examples of *gene families* comprising several distinct but related sequences, although every member probably occurs in at least a few identical copies. A similar situation is found with many single-copy genes, which may also occur in families of related sequences. For instance the globin gene family in man comprises three inactive (pseudogene) and eight active members, of which only two (coding for α globin) are virtually identical. The rest represent divergent members of a sequence family that has undergone duplication several times during evolution. The organisation and sequential expression of globin genes will be considered in detail in chapter 5.

Smaller families of related genes are known in the cases of chick ovalbumin and *Xenopus* vitellogenin (chapter 6).

Many structural proteins are represented in the genome by gene families with variable numbers of members, e.g. those coding for keratins (Kemp, 1975), collagens (see Tolstoshev & Solomon, 1982), tubulins (Cleveland, 1983), actins etc. In these examples, several discrete gene products – each with distinctive functions and/or cellular locations – are encoded by the different members of the family. Certain members of the actin gene family are expressed at low levels in a wide variety of cell types; these actins play a key role as contractile elements in the cytoskeleton. Related but not identical *isoforms* of actin are expressed at much higher levels in skeletal or cardiac muscle tissue, whose primary function is contractile. There are also developmental changes in actin type between foetal and adult stages. The same species of actin mRNA is expressed in both skeletal and cardiac muscle in the mouse foetus, but in adults it is confined to cardiac muscle; a different type of actin gene is presumably activated in adult skeletal muscle (Minty *et al.*, 1982). The actin genes of *Drosophila* form a family of about six members dispersed to several chromosomal sites (Fyrberg *et al.*, 1980), and each of these genes is expressed in a stage- and tissue-specific manner (Fyrberg *et al.*, 1983). A similar heterogeneity has been reported for the actin genes in animals ranging from sea urchins to mammals. However, structural proteins are not all derived from such gene families; the cytoskeletal protein vimentin, for instance, is encoded by a single gene in the hamster genome and in other higher vertebrates (Dodemont *et al.*, 1982; Quax *et al.*, 1983). However, transcripts from the single chick vimentin gene can be processed by two alternative pathways to yield messenger RNA species of 2.2 and 2.5 kb, apparently resulting from the use of two different poly(A) addition sites (see §3.7; Zehner & Paterson, 1983).

Several enzymes are also known to occur in alternative forms in different tissues. At the gene level, either (a) closely related genes may be expressed in a tissue-specific manner, or else (b) transcripts from a single gene may be processed differentially (see §3.9) to generate non-identical mRNAs (e.g. for rat pyruvate kinase L and L′ subunits; Marie *et al.*, 1981). A more complex situation is found in the mouse α amylase system described by Schibler *et al.*, (1982, 1983). Two copies of the *Amy-2^a* gene specify the pancreatic form of α amylase, while the liver and salivary-gland form of the enzyme is encoded by a single *Amy-1^a* gene (closely related to *Amy-2^a*). The vastly different levels of α amylase activity in liver (low) and in salivary gland (high) result from tissue-specific utilisation of two different promoter sites (§3.4) for the initiation of *Amy-1^a*

transcription; the strong promoter is used only in salivary gland, while the weak promoter is used in both tissues.

1.10 Other features of animal DNA

Consideration of the structure of single-copy protein-coding genes is deferred until chapter 3, since their putative promoter sequences and the existence within them of introns (non-coding DNA sequences interrupting the coding region) have a close bearing on transcription and RNA processing. Certain well-studied genes belonging to this class (globins, ovalbumin, vitellogenin) are dealt with in more detail in the second part of this book (chapters 5 and 6).

Many different sequences are represented in the single-copy fraction of eucaryotic DNA, and this makes it difficult to draw general conclusions about their organisation beyond the interspersion patterns described earlier (§1.6). Long domains containing up to 100 kbp of DNA may underlie certain features of higher-order chromatin packing (§2.4) and the phenomenon of lampbrush chromosome loops during oogenesis (§4.4). Such domains may possibly delineate groups of functionally related genes; stretches of AT-rich sequences have recently been reported to frame such groupings, e.g. the two clusters containing the α- and β-globin-related genes respectively (Moreau *et al.*, 1982).

Finally, four specialised features of eucaryotic DNA should be mentioned briefly.

1.10.1 Transposable genetic elements

These are sequences of DNA up to several kilobase pairs in length, which can move from one site to another in the genome. Among animals, they have been studied most extensively in *Drosophila*, where *in situ* hybridisation to polytene chromosomes clearly demonstrates their mobility (different locations in different strains). Several transposable elements such as *copia*, 412 and 297 have been characterised as distinct repetitive-sequence families whose members are dispersed to numerous (but variable) sites within the *Drosophila* genome (Potter *et al.*, 1979). Transposable elements are characteristically flanked by direct repeats of 200–500 bp, a feature also shared by retrovirus genomes. Retroviruses are single-stranded RNA viruses, which can become inserted into host-cell chromosomes in the form of duplex DNA proviral copies (§3.2).

Copia sequences are abundantly transcribed into messenger-like cytoplasmic RNAs, which are found packaged into discrete retrovirus-like particles in *Drosophila* cells (Shiba & Saigo, 1983). Moreover, Flavell & Ish-Horowicz (1981) have discovered extrachromosomal circular copies of the *copia* DNA sequence, which resemble the circular intermediates in proviral insertion and excision. This evidence suggests that *copia* and other transposable elements may be related to the chromosomal provirus form of retroviruses, implying a common evolutionary origin (see Temin, 1980; Flavell & Ish-Horowicz, 1983). One consequence of the mobility of transposable elements is that mutations frequently result at their sites of insertion. For instance, a 7.3 kbp transposable element known as HMS Beagle is present in about 50 dispersed copies in the *Drosophila* genome; in one particular mutant strain, this element is found splitting the putative promoter site of the CP3 larval cuticle protein gene, hence inactivating it (Snyder *et al.*, 1982). Other examples of transposable-element insertions causing mutations, e.g. in the *Drosophila* bithorax complex, will be mentioned in chapter 7.

1.10.2 Deletion and amplification of DNA sequences

Examples of both processes are known, and are summarised in section 4.2 in the context of genome constancy and nuclear totipotency. The special case of gene rearrangements during the differentiation of antibody-producing cells is also considered briefly in section 4.2. The comparative rarity of gene amplification, deletion and rearrangement processes during development implies that the genome remains constant in most differentiating animal cells, at least on a gross level (see e.g. Levine *et al.*, 1981).

1.10.3 Z-form DNA

The vast majority of duplex DNA occurs in the classical B form, with two anti-parallel strands hydrogen-bonded together in a right-handed helix. However, certain synthetic polydeoxyribonucleotides, particularly alternating purine-pyrimidine copolymers such as poly(dGdC).poly(dCdG), adopt an alternative Z-form helix which is left-handed. Antibodies directed against Z-form DNA (Lafer *et al.*, 1981) recognise sequences within *Drosophila* chromosomes (Nordheim *et al.*, 1981; Arndt-Jovin *et al.*, 1983), although such Z-DNA regions are only apparent after acid fixation of the chromosomes (Hill & Stollar, 1983). A number of *Drosophila* proteins which bind specifically to Z-form DNA have been isolated (Nordheim *et al.*, 1982). Possible signalling roles for permanent or transient Z-form sequences in the chromosome remain for the time being speculative.

1.10.4 Mitochondrial genomes

So far this discussion has been confined to nuclear DNA sequences. However, the mitochondria in animal cells contain small DNA circles (mDNA) encoding a small range of mitochondrion-confined RNA and protein products. These include two rRNA species which are more closely related in size and sequence to bacterial 16S and 23S rRNAs than to the 18S and 28S rRNAs encoded by the nuclear genes of the same animal (Küntzel & Kochel, 1981). Some but by no means all of the characteristic mitochondrial proteins (e.g. enzymes of the electron transport chain) are also encoded by mDNA; the others are coded by nuclear genes. In mammalian mitochondria, the ribosomal gene region is represented in two overlapping transcription units read from separate promoters on the same mDNA strand (Montoya *et al.*, 1983); one yields the rRNA precursor, while the other gives rise to a polycistronic RNA molecule from which the mitochondrial mRNAs and most tRNAs are derived.

Eucaryotic mitochondria may have evolved from symbiotic procaryotic ancestors, but if so a number of mitochondrial functions have been lost from the mDNA and taken over by nuclear genes. Whether actual genes have been transferred is more debatable, though rearranged mitochondrial sequences have recently been discovered in the nuclear DNA of yeast (Farrelly & Butow, 1983). Fungal mitochondrial genomes are much larger than their animal counterparts, perhaps implying an independent procaryotic origin (supported by the available rRNA sequence data; Küntzel & Kochel, 1981).

2

Chromatin

Summary

Eucaryotic chromosomes are packaged within the interphase nucleus in the form of *chromatin*, a complex composed mainly of protein and DNA together with some RNA (§2.1). Two major classes of protein contribute towards chromatin structure, namely histones and non-histone proteins. There are five major types of histone (H1, H2A, H2B, H3 and H4), though some of these occur in multiple subtypes (§2.2). Additional variety can be generated by post-translational modifications, including reversible acetylation and phosphorylation processes (§2.2). Non-histone proteins are much more heterogeneous, and defined roles have been assigned to only a few members of this class (§2.2).

The fundamental repeating unit of chromatin organisation is a bead-like structure known as the *nucleosome* (§2.3). Within this structure, an octamer of histones ($H2A_2 \cdot H2B_2 . H3_2 \cdot H4_2$) protects 145 bp of DNA irrespective of sequence. This *core unit* repeats once every ~200 bp along the length of the chromosomal DNA; one molecule of H1 histone is associated with the ~60 bp of *linker* DNA that separate neighbouring core units. The major structural features of the nucleosome are determined by the association between core DNA and an $H3_2 \cdot H4_2$ tetramer, perhaps accounting for the extreme evolutionary conservatism of these two histones (§2.3).

Nucleosomal chromatin is generally coiled into a solenoid structure of diameter 30–35 nm. This compacts the overall DNA length by some 40 fold, i.e. 6–7 fold within the nucleosome (where the core DNA is wound around the outside of the histone octamer), and a further 6–7 fold within the solenoid (§2.4). In condensed metaphase chromosomes, DNA packing ratios of several thousand fold are required. This is achieved by organising the chromosomal DNA into domains or loops which are anchored at their bases by a scaffolding of non-histone proteins (§2.4).

Actively transcribed regions of DNA are present in an altered chromatin conformation, known as smooth-fibre chromatin from its non-beaded appearance in EM (§2.5). Here the DNA length is compacted by only two fold, yet the fundamental 200 bp periodicity and core histone complement characteristic of nucleosomal chromatin are retained within the smooth-fibre regions (§2.5). Such smooth-fibre chromatin is preferentially digested by the enzyme DNase I, and perhaps by other nucleases. The DNase I-sensitive chromatin regions include not only active genes, but also some genes which have been or will become active in the cell type under study, as well as the DNA sequences flanking such genes (§2.5). DNase I-sensitive chromatin is distinguished from nucleosomal chromatin by the presence of two specific non-histone proteins designated HMGs 14 and 17 (§2.6). DNase I-hypersensitive sites are commonly found close to the 5' ends of actively transcribed or potentially transcribable genes (§2.6). The DNA sequences of 'active' gene regions (or at least their 5'-flanking sequences) are generally undermethylated as compared with bulk nucleosomal DNA (highly methylated) in vertebrates; however, undermethylation is not always prerequisite for transcription to occur. Notably, the patterns of DNA methylation and of DNase I-hypersensitive sites are heritable (§2.6).

2.1 Introduction

Apart from their contrasting modes of organisation, eucaryotic and procaryotic DNAs are also distinguished by their packaging. The bacterial chromosome is somewhat thinly coated with proteins; some are involved in folding the DNA, while others include the enzymes required for replication, recombination and transcription. Eucaryotic chromosomes, by contrast, are thoroughly cocooned in proteins whose total mass greatly exceeds that of the DNA (by around two to one). Principal among these proteins are the five major histones (see §1.9), together with an ill-defined mixture of non-histone proteins. The latter are much more heterogeneous than the histones, and include many enzymes involved in the synthesis and processing of RNA and DNA, as well as structural components of the chromosome. The entire DNA : histone : non-histone-protein complex, together with some associated RNA, is known as *chromatin*; it represents the decondensed interphase form of the nuclear chromosomes.

Whereas the skimpily-clad bacterial chromosome mingles and interacts freely with its cellular environment, this is not the case for its heavily mantled eucaryotic counterparts. The presence of the nuclear membrane in eucaryotes creates a kind of harem, with strict control of access to and exit from the nuclear compartment. A recent review by de Robertis (1983) shows elegantly how particular protein and RNA species partition themselves between nucleus and cytoplasm. Another effect of the nuclear membrane is to separate *transcription* (DNA to RNA) from *translation* (RNA to protein). Though intimately associated in bacterial cells, these processes are quite distinct in eucaryotes, the former being confined to the nucleus while the latter is exclusively cytoplasmic (mitochondria and plant chloroplasts are exceptional on both counts).

This introduces two additional levels at which gene expression can be controlled in higher organisms, namely RNA processing and RNA export from nucleus to cytoplasm (presumably via the nuclear pores). The first of these is dealt with in the next chapter, while the second remains largely unexplored – though it is clearly of some importance to our understanding of gene control. Thus when seeking features which distinguish actively transcribed from non-transcribed regions of eucaryotic chromatin (a major topic of this chapter), we should bear in mind that this is only the first level at which gene expression might be regulated.

2.2 Chromatin proteins

The five major histone proteins (H1, H2A, H2B, H3 and H4) are almost universally associated with the nuclear DNA in animals. One exception to this generalisation is found in the spermatozoa of many fish (e.g. trout); during the maturation of these cells, the histones are progressively replaced by small, basic, arginine-rich proteins known as protamines. In other animal groups, sperm-specific histones may fulfil a similar role.

All five major histones contain an unusually high proportion of basic amino acids; H1 is very rich in lysine, H2A and H2B are moderately rich in lysine, while H3 and H4 are both rich in arginine. Since DNA is acidic (in effect an enormously long polyanion), close interactions with these basic residues on the histones would be expected. A sixth histone designated H5 has been found in addition to H1 in the transcriptionally inactive chromatin of nucleated erythrocytes in birds and amphibia. H5 is also very rich in lysine, but is relatively richer in arginine than H1.

The ease with which histones can be removed from the DNA by increasing the ambient salt concentration is $H1 > H2A \simeq H2B > H3 \geqslant H4$. The same order is obtained by comparing the extent of sequence deviation between corresponding histones from a wide range of eucaryotes. Thus H1 varies considerably in sequence between different organisms, particularly in the extended N- and C-terminal portions of the molecule (the central globular portion is better conserved). H2A and H2B are much less variable, and most of the amino-acid substitutions observed between species are confined to the extended N-terminal parts of both molecules. H3 and H4 are the most extreme examples of evolutionary conservatism known among eucaryotic proteins; between mammals (cow) and higher plants (pea) there are only two amino-acid changes in H4 and four in H3. The fact that H3 and H4 are also very difficult to dissociate from the DNA using high salt, suggests a tight interaction which may impose severe constraints on the extent of permissible sequence variation.

H2A, H2B, H3 and H4 are found in equimolar proportions in chromatin prepared from diverse eucaryotic sources, while H1 is present in approximately half the molar amount of the other histones. A DNA-histone complex of the general form $(H2A_2 \cdot H2B_2 \cdot H3_2 \cdot H4_2 \cdot H1)$: DNA is implied by this; its structure is considered below in section 2.3 on the nucleosome.

Before leaving the histones, some mention should be made of their variant forms, including modifications to some of their amino-acid residues. The existence of different subsets of H1, H2A and H2B histone genes has already been noted in section 1.9. The fact that they are sequentially expressed during development suggests some functional role for these variants. Simpson (1981) has observed that sea urchin nucleosomes (§2.3) containing different H1 subtypes show different stabilities *in vitro*, which might be reflected *in vivo* in properties such as replication rate or conversion to the transcriptionally active conformation (§2.5).

Even within the same histone subtype, considerable variety can be engendered through several post-translational modification systems, including methylation, acetylation and phosphorylation of specific residues. Thus certain lysine side-chains in histones H3 and H4 can become methylated irreversibly, forming mono-, di- or trimethyl lysine derivatives. The α-NH_2 group of the N-terminal serine in H1, H2A and H4 may undergo irreversible acetylation, whereas the ϵ-NH_2 groups of lysine residues in the N-terminal regions of H2A, H2B, H3 and H4 are available for reversible acetylation. This latter modification neutralises the net positive charge on the lysine side-chain (as acetyllysine), so reducing the basicity of those N-terminal regions affected. Some correlation has been claimed between the prevalence of hyperacetylated H4 (with three or four acetyllysines) and the transcriptional activity of chromatin; however, recent evidence from *Physarum* seems to argue against such a link (Loidl *et al.*, 1983).

Reversible phosphorylation affecting serine and threonine side-chains (converting them from neutral to negatively charged residues) occurs in all types of histone. H1 is particularly susceptible, with up to seven sites becoming phosphorylated. For the most part, H1 phosphorylation occurs in phase with the cell cycle; thus fast-growing cells contain substantially more phosphorylated H1 than do non-dividing cells. H1 phosphorylation takes place mainly during the G2 phase of the cell cycle, e.g. in the slime-mould *Physarum* (Bradbury *et al.*, 1974a, b), and is associated with the first stages of chromosome condensation in mitotic prophase. It is suggested that phosphorylated H1 favours a closer packing of nucleosomes (§2.3) or higher-order chromatin structures (§2.4). By contrast, the dephosphorylation of H1 which occurs during and after metaphase (Balhorn *et al.*, 1975), would allow the chromatin to adopt a less condensed configuration during the subsequent interphase. Another type of histone phosphorylation is specific to the serine residue at position 37 in H1, and is mediated by a cyclic-AMP-dependent protein kinase. Intracellular cyclic AMP (cAMP) levels increase in response to a variety of hormones which bind to cell-surface receptor proteins and thereby activate the cAMP-producing enzyme adenyl cyclase on the cytoplasmic face of the membrane. The fact that H1 histone is one of the target sites for cAMP-dependent protein kinase, suggests one possible mechanism whereby a hormone acting at the cell surface could alter the chromatin structure and transcriptional activity of the nucleus.

Other histone modifications include poly ADP-ribosylation and the formation of protein A 24. The latter is a bifurcated protein, formed by covalent linkage of a globular non-histone protein (ubiquitin) to about 10% of H2A molecules via the side-chain ϵ-NH_2 group of the lysine residue at H2A position 119. A similar linkage of ubiquitin to H2B has been described more recently (West & Bonner, 1982).

As mentioned in the introduction (§2.1), the non-histone proteins of chromatin are a heterogeneous and ill-defined group. They include: (a) DNA and RNA polymerases and processing enzymes; (b) proteins contributing to chromosome structure (e.g. metaphase scaffolding proteins; see §2.4); (c) proteins affecting chromatin conformation (e.g. two members of the high mobility group, HMGs 14 and 17; see §2.5); and (d) putative gene regulatory factors (e.g. TFIIIA, see §3.4; also steroid-receptor complexes, see §4.8 and chapter 6). We will return to these examples later, as indicated. Both tissue- and stage-specific changes affect the pattern of non-histone proteins, as seen when these are extracted from nuclei and separated by gel electrophoresis. But in the absence of identified functions for those components showing such changes, it is difficult to assess their significance.

2.3 The nucleosome

The basic organisation of DNA and histones in chromatin was clarified only in 1973–5, when several lines of investigation came to fruition almost simultaneously; namely electron microscopy (EM) of chromatin, identification of histone complexes from chromatin, reassembly of chromatin from purified DNA and histone components, and nuclease digestion of chromatin into discrete subunits.

The first EM visualisation of chromatin as a beaded structure (Olins & Olins, 1974) was achieved by lysing interphase nuclei into water and centrifuging the released material onto grids following fixation. This revealed chromatin fibres escaping from the ruptured nuclei in the form of 'beaded strings' with particles 6–8 nm in diameter (called *nu* bodies) arranged at regular intervals along a much narrower (1.5 nm) filament. To some extent the dimensions of these structures were reduced by the fixation method used; presently accepted values for *nu*-body diameter are in the range 11–13 nm with unfixed material. Isolated chromatin shows a similar beaded structure under EM if it is first depleted in histone H1 by mild salt or trypsin treatment (Oudet *et al.*, 1975), suggesting that H1 is not an essential component of *nu*-body structure. When the full complement of H1 is present, however, isolated chromatin tends to form aggregates, obscuring the underlying nucleosome pattern when viewed under EM.

Dissociation of histones from chromatin under mild conditions led to the identification of histone multimers, in particular a tetrameric $H4_2 \cdot H3_2$ complex (Kornberg & Thomas, 1974). This in turn led to the first enunciation of the nucleosome model for chromatin structure (Kornberg, 1974). Histone reassociation studies confirmed that equimolar amounts of H3 and H4 can form tetramers ($H3_2 \cdot H4_2$) spontaneously in solution; furthermore, this tetramer structure can bind two molecules each of H2A and H2B in the presence but not in the absence of DNA. This generates a DNA-associated histone octamer ($H2A_2 \cdot H2B_2 \cdot H3_2 \cdot H4_2$) which repeats periodically along the length of any type of DNA duplex, independent of sequence (see e.g. Oudet *et al.*, 1975, 1978). The form and dimensions of these histone-octamer:DNA complexes, as revealed by EM, are very similar to those of the *nu* bodies found in interphase chromatin. The term *nucleosome* has been generally adopted for this basic repeating unit of chromatin structure, and will be used from here on in preference to synonyms such as *nu* body.

It had already been shown by Hewish & Burgoyne in 1973 that rat liver chromatin contains an endogenous nuclease which, when activated by Ca^{2+} ions, cleaves the DNA component into a series of fragments corresponding to monomers and multimers of a fundamental 200 bp unit. This is seen clearly as a 'ladder' of fragment sizes after agarose gel electrophoresis of the DNA. Subsequently these studies were extended by Noll (1974a) and others, using an exogenous enzyme such as micrococcal nuclease to digest the chromatin (fig. 18). Since chromatin from virtually any source is cleaved by micrococcal nuclease into ~200 bp fragments, while deproteinised DNA is cleaved essentially randomly, the 200 bp repeat must reflect some feature of chromatin structure rather than DNA sequence. Monomer, dimer, trimer and tetramer fragments of chromatin released by micrococcal nuclease were each purified from a sucrose density-gradient and examined under EM; as expected, they corresponded to single nucleosomes and to groups of two, three or four linked nucleosomes, respectively (Finch *et al.*, 1975).

There was initially some controversy as to the monomer size of the DNA fragments prepared from micrococcal nuclease-digested chromatin; some estimates gave 145 bp while others ranged around 200 bp. It later transpired that these disparities were due to differences in the methodology used. At 0°C small amounts of the nuclease produce double-stranded cuts with a spacing of approximately 200 bp or multiples thereof; these cuts occur at sites located in the so-called *linker* DNA regions between the nucleosome core units. This generates a ladder of DNA fragments with sizes of ~200 (monomer), 400, 600, 800 etc. base pairs when separated on agarose gels. At higher temperatures (e.g. 37 °C) or with larger amounts of nuclease, the DNA of the linker regions is digested away, leaving a monomer *core* particle in which 145 bp of DNA (termed core DNA) is protected by the histone octamer from further attack. Nucleosome multimers released under these conditions will of course lose linker DNA only from their ends, leaving uncleaved linker regions joining adjacent nucleosome cores within the fragment. This generates a series of DNA fragments of approximate sizes 145 (monomer), 345, 545, 745 etc. base pairs, as shown diagrammatically in fig. 18 (for details, see Kornberg, 1977).

In fact, the 200 bp repeat length in chromatin is subject to considerable

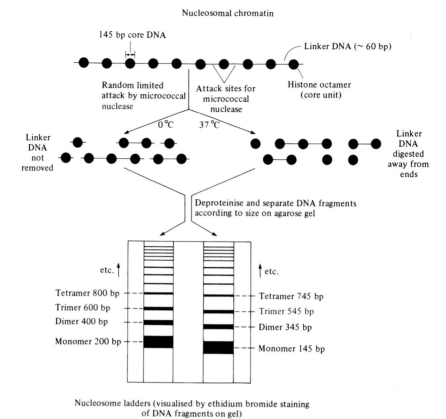

Fig. 18 Micrococcal nuclease digestion of chromatin.

variation, ranging from 160–170 bp in mammalian cortical neurons up to 241 bp in sperm cells from the sea urchin *Arbacia*. In gastrula cells from *Arbacia* embryos, the chromatin repeat length is about 220 bp, considerably shorter than in sperm cells from the same species (Spadafora *et al.*, 1976). The short repeat length in mammalian cortical neurons is of particular interest, since glial-cell chromatin from the same tissue has a longer repeat length close to 200 bp (see Thomas & Thomson, 1977). Cerebellar neurons, by contrast, have a long repeat length of 218 bp. The changes in chromatin repeat length which occur in cortical (200 → 170 bp) and cerebellar (165 → 218 bp) neurons have been shown to accompany terminal differentiation of these cells (Jaeger & Kuenzle, 1982). However, in all cases the core DNA length protected by the histone octamer remains constant at 145 bp; thus variations in overall repeat length result exclusively from different lengths of linker DNA (see Kornberg, 1977).

The basic structure of the nucleosome which emerges from these studies is as follows. In the core particle, 145 bp of DNA (irrespective of sequence) are associated directly with a histone octamer comprising two molecules each of H2A, H2B, H3 and H4. A single molecule of the remaining histone, H1, is associated with each stretch of linker DNA between adjacent nucleosome cores. Evidence for this is provided by subjecting whole nucleosomes to further micrococcal nuclease digestion to produce core particles. The H1 histone initially present is lost during the 'trimming' of linker DNA, while the content of the other four 'core histones' remains constant (Noll & Kornberg, 1977). Some evidence suggests that H1 is located on the linker close to the point where the DNA strand leaves the nucleosome core particle. This would account for the formation of 168 bp intermediates still containing H1, during digestion of whole nucleosomes down to core particles (Noll & Kornberg, 1977). In tissues expressing H5 histone (e.g. nucleated avian erythrocytes), one molecule of H5 is added to each linker during differentiation, apparently without displacing the H1 already present (Weintraub, 1978). Thus two very lysine-rich histone molecules (H1 and H5) become bound to each linker, and indeed the linker length also increases from 58 to 72 bp during avian erythroid differentiation (Weintraub, 1978).

It remains to describe the internal structure of the nucleosome, i.e. how the DNA and histones are arranged in each nucleosome particle. The DNA within core particles (145 bp unit) is susceptible to endonucleolytic attack by several nucleases, including DNases I and II, which introduce staggered cuts at approximately 10 bp intervals or multiples thereof. This spacing is consistent with one attack site per turn of the DNA double helix (Noll, 1974b; Sollner-Webb *et al.*, 1978), and suggests that the core DNA is located at or close to the surface of the histone octamer in order to allow access by the nuclease. The arguments involved are actually much more complex than this, and the reader is referred to chapter 13 of Lewin (1980) for a fuller discussion. Biophysical data also suggest that the DNA forms an outer shell around a smaller protein core within the nucleosome, and further indicate that the nucleosome is disc-shaped rather than spherical. The DNA is wound around the outside of the histone octamer, probably giving 1.7 turns per core particle. This helical packing of the DNA can account for most of the 6–7 fold length reduction observed e.g. when fragments of duplex DNA are packaged into core particles *in vitro* through association with histones H2A, H2B, H3 and H4 (Oudet *et al.*, 1975).

As to the arrangement of histones within the nucleosome, a subnucleosomal particle can be generated from DNA and histones H3 + H4 alone, forming an $H3_2 \cdot H4_2$ tetramer protecting 145 bp of DNA. The nuclease susceptibility and X-ray diffraction patterns of these particles are very similar to those of intact nucleosomes, suggesting that the major structural features of the nucleosome are determined by the interactions of H3 and H4 with DNA. This is consistent with the extreme evolutionary

A   Spatial organisation

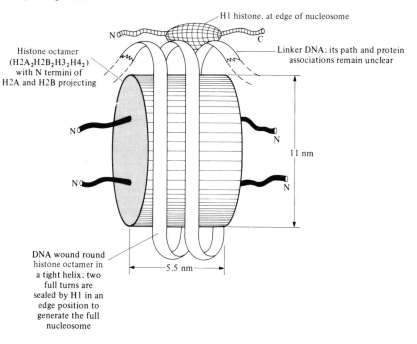

B   Summary of structural parameters

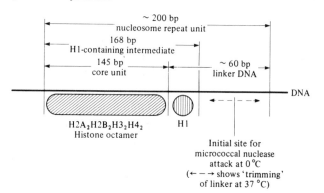

Fig. 19 Model of nucleosome structure (part *A* modified from Bradbury *et al.*, 1981).

conservatism of H3 and H4 (see §2.2). The roles of H2A and H2B are less clearly defined. Their C-termini and globular central portions (i.e. the most conserved regions) interact directly with each other, with H4 and with DNA in the nucleosome core, perhaps stabilising its structure. Agents such as UV light and tetranitromethane can be used to cross-link proteins together at points of close contact; in nucleosomes, such contacts are identified particularly between H2B and H4, and between H2A and H2B. The two H3 molecules within each core can also be cross-linked via disulphide-bridge formation between two cysteine residues, and there is evidence of other (less close) contacts e.g. between H2A and H3. The variable N-terminal regions of H2A and H2B do not apparently contribute to core particle structure, and may be available to interact with linker DNA. The conserved central globular region of H1 is probably bound at the edge of the core particle disc, where it could seal off the nucleosome structure, so completing two full turns of the DNA around each histone core (only 1.7 turns are completed in core particles lacking H1). The variable N- and C-terminal regions of H1 probably play an important role in higher-order packing of the chromatin (see §§2.2 and 2.4). This model is shown diagrammatically in fig. 19.

2.4 Higher order structure of chromatin

As mentioned in the previous section, nucleosome formation results in a 6–7 fold compaction of the extended DNA length. However, the average mammalian cell contains well over a metre of DNA, and this must somehow be packaged inside a nucleus only ~5 μm in diameter! A DNA packing-ratio greater than seven would seem likely even for interphase chromatin, while the known lengths of metaphase chromosomes imply a packing-ratio of many thousands. EM studies of isolated interphase chromatin show two major types of fibrous structure distinguished by their diameters of 10–11 nm and 30–35 nm respectively.

The thinner *nucleofilament* form is prevalent at low ionic strengths (<20 mM NaCl) and consists of a linear chain of nucleosomes (11 nm diam.), perhaps arranged edge to edge. When stretched slightly this gives the familiar 'beads on a string' pattern (alternating nucleosome cores and thinner linker regions). There is no significant packing within this structure beyond the 6–7 fold already achieved in the component nucleosomes. As the ionic strength is increased, particularly with respect to divalent cations, so the linear nucleofilament structure undergoes a two-stage transition to a thicker (30–35 nm diameter) 'supercoil' form, as shown diagrammatically in fig. 20.

Low ionic strength (< 20 mM NaCl)   Intermediate ionic strength (> 20 mM NaCl)   High ionic strength or divalent cations (200–300 mM NaCl or 2 mM MgCl$_2$)

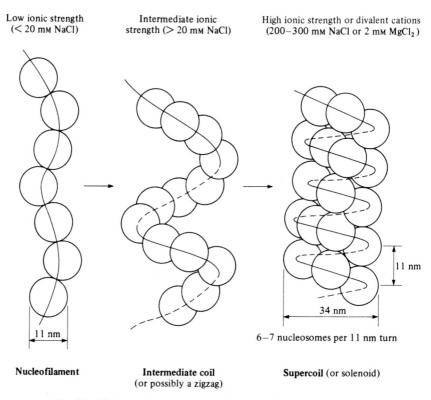

11 nm

34 nm

6–7 nucleosomes per 11 nm turn

Nucleofilament   **Intermediate coil** (or possibly a zigzag)   **Supercoil** (or solenoid)

Fig. 20 Higher order structure of chromatin (modified from Bradbury *et al.*, 1981).

The supercoiled fibre (or *solenoid*) achieves a further compaction of 6–7 fold over and above the nucleosomal level, i.e. an overall packing-ratio of 40 fold compared with extended DNA. It is probable that the supercoil form is stabilised by interactions involving the variable N- and C-terminal regions of H1 histone, which are subject to cyclic phosphorylation and dephosphorylation during the cell cycle (see §2.2). Thus H1 is a bifunctional protein, since it also seals off the nucleosome structure through its globular central region (conserved). The 30–35 nm solenoid is probably the major structural form present in interphase chromatin, as suggested by EM and by X-ray and neutron diffraction studies.

Packing-ratios of the order of several thousand are required in order to compact a metre of DNA down to a hundred microns or so, representing the summed lengths of all the metaphase chromosomes in a cell about to divide. There is no reason in principle why a chromatin solenoid could not be wound into a hierarchical series of super-supercoils (etc.) in order to achieve the required degree of condensation. However, the elegant EM

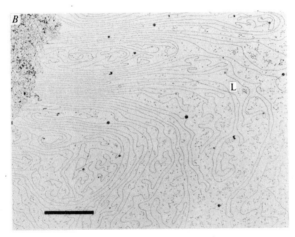

Fig. 21 Electron micrographs showing the metaphase chromosome 'scaffolding' (S, part *A*) of non-histone proteins anchoring many large loops (L, part *B*) of histone-depleted DNA. Bar represents 1μm. Photographs reprinted with permission from the copyright holders, Cold Spring Harbor Press. From U.K. Laemmli *et al.* (1977) *Cold Spring Harbor Symp. Quant. Biol.* **42**, 351–60.

studies of Laemmli and coworkers (1977; see fig. 21) suggest a more radical solution to the problem.

In brief, metaphase chromosomes were found to maintain a relatively condensed structure even when entirely depleted of histones by extraction with 2 M NaCl (example in fig. 21). The DNA is found organised into numerous loops or *domains* ranging between 35 and 70 kbp in length; these radiate outwards like a halo around a central matrix or scaffolding structure, which corresponds in size and shape to the original metaphase chromosome. The scaffolding is itself composed of some 20 different non-histone proteins, which apparently hold the entering and leaving DNA strands in place at the base of each loop.

2.5 Transcriptionally active chromatin

In 1975 came the discovery that certain regions of interphase chromatin are devoid of beaded nucleosomes, instead taking the form of thinner smooth fibres (see e.g. Oudet *et al.*, 1975). It was suggested originally that these regions might represent histone-free chromatin or even naked DNA, and further that they might be actively engaged in transcription. However, the diameter of smooth-fibre regions is some 3.5 nm, as against 2 nm or less for naked DNA. It is now thought that the basic nucleosome structure is retained in smooth-fibre chromatin, but in a more extended conformation which compacts the DNA length by only 2 fold, rather than 6–7 fold as in nucleosomes.

Several independent lines of evidence support this conclusion. For example, antibodies directed against H2B and H3 histones recognise smooth-fibre regions of chromatin as well as nucleosomes, i.e. these histones at least must be present in the extended smooth-fibre structure (McKnight *et al.*, 1977). In *Physarum* rDNA chromatin, all four core histones are present even when these genes are active. However, H3 can be labelled by a fluorescent sulphydryl reagent only in transcribed regions, but not in non-transcribed regions, nor when the ribosomal genes are inactive (Prior *et al.*, 1983). This indicates a structural difference between the extended conformation of nucleosomes in actively tran-scribed regions and the compact conformation in non-transcribed regions; note that this distinction affects the accessibility but not the presence of H3.

It is also known that the 200 bp DNA fragments released by micrococ-cal nuclease digestion of chromatin include sequences complementary to mRNAs expressed abundantly in the tissue under study (e.g. globin gene sequences from erythroid chromatin, or ovalbumin gene sequences from oviduct chromatin). This implies that the 200 bp periodicity and nuclease-sensitivity of nucleosomes applies equally to actively transcribed chromatin. A similar conclusion emerges from psoralen cross-linking studies. The two strands of duplex DNA can be cross-linked covalently with trimethylpsoralen, and this technique is applicable to chromatin *in situ* within the cell nucleus. When chromatin is extracted from psoralen-treated *Drosophila* nuclei and the DNA component is then denatured and examined by EM, almost all of the strands are found to be cross-linked at 200 bp or longer intervals (fig. 22), with very little DNA cross-linked at shorter intervals (Cech *et al.*, 1977). Thus the cross-linking agent, like micrococcal nuclease, can only gain access to the DNA at one site per nucleosome unit, and this basic pattern must persist in smooth-fibre as well as beaded chromatin (fig. 22).

Definitive evidence for a different chromatin conformation at active as opposed to inactive gene sites is provided by the greater sensitivity of the former to certain endonuclease enzymes, particularly DNase I. Under some conditions both DNase II and micrococcal nuclease are reported to attack active (smooth-fibre) chromatin in preference to bulk nucleosomal chromatin, but these results have proved less easy to reproduce than those obtained using DNase I. In general, the chromatin at active gene sites (i.e. those known to be transcribed *in vivo*) is some 25 times more sensitive to DNase I attack than is the chromatin at non-transcribed sites. Thus for example, the globin gene regions are DNase I-sensitive in erythroid chromatin (where they are actively transcribed), but are much less sensitive in oviduct chromatin (where they are never transcribed).

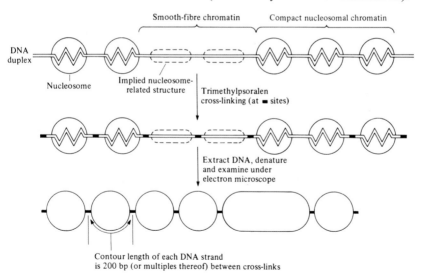

Fig. 22 Psoralen cross-linking of DNA strands in chromatin.

Under conditions where only 10% of the bulk DNA is digested by DNase I, some 70% of the globin sequences are lost from erythroid chromatin. Similarly, the DNase I-sensitivity of the ovalbumin gene region is much greater in oviduct than in erythroid chromatin. However, some caution is needed when interpreting such results in terms of a direct relationship between gene activity and the DNase I-sensitivity of that site in the chromatin. Three main reasons are given below.

(i) The DNase I-sensitive sites identified in chromatin include some previously active and/or potentially active genes as well as genes presently engaged in transcription. One example of a previously active gene remaining DNase I-sensitive after cessation of active transcription, is the foetal γ globin gene in adult sheep erythroid chromatin (Young *et al.*, 1978). An opposite example is found in early chick-embryo erythroid chromatin, where the adult β globin gene is already DNase I-sensitive, although it is not transcribed until later in development (Stalder *et al.*, 1980a). In hen oviduct chromatin, where ovalbumin expression is under hormonal control, the ovalbumin gene site remains DNase I-sensitive during hormone withdrawal, even though transcription ceases (Palmiter *et al.*, 1977). These examples will be discussed in more detail in part 2 of this book (see chapters 5 and 6).

(ii) The DNase I-sensitive region of chromatin may extend far beyond the transcribed gene itself. In chick erythroid cells, a chromatin region extending for 8 kbp on the 3′ side and 6–7 kbp on the 5′ side of the β globin genes is found to be DNase I-sensitive (Stalder *et al.*, 1980b). Similarly, a region of at least 30 kbp around the ovalbumin gene is DNase I-sensitive in oviduct chromatin. This region includes two genes (X and Y) related to the ovalbumin gene and expressed under similar hormonal control, but also contains non-transcribed regions (Lawson *et al.*, 1980).

(iii) Finally, there are varying degrees of DNase I-sensitivity. Rarely transcribed gene sites in chromatin appear slightly less DNase I-sensitive than abundantly transcribed gene sites (see Garel *et al.*, 1977). Also, the DNase I-sensitive chromatin regions flanking the chick α and β globin genes are not digested as rapidly as the chromatin of the genes themselves (Stalder *et al.*, 1980b); the former should properly be described as moderately DNase I-sensitive, while the latter are highly DNase I-sensitive. In addition, there are DNase I-hypersensitive sites (see §2.6) in chromatin regions close to the 5′ ends of many actively transcribed or potentially transcribable genes.

In summary, these DNase I-digestion studies identify a sensitive component of chromatin comprising decondensed 'active' regions. This contrasts with the relative DNase I-resistance of bulk chromatin, where non-transcribed DNA sequences are packaged into compact nucleo-somes. Electron microscopy reveals regions of 3.5 nm smooth-fibre chromatin, often visibly engaged in transcription when studied by EM-spreading techniques; these regions apparently represent the DNase I-sensitive component. Despite its non-beaded appearance, the smooth-fibre material retains the 200 bp periodicity and core histone complement of beaded nucleosomal chromatin; this presumably results from a more extended conformation of the basic nucleosome unit, with 2 fold rather than 6–7 fold compaction of the DNA length.

However, the DNase I-sensitive conformation does not uniquely identify those genes engaged in transcription. In addition, genes which were active previously, or which will become active at a later stage, as well as extensive flanking regions around such genes, may all be present in a DNase I-sensitive form. Genes transcribed at very different rates are distinguished by only slight differences in DNase I-sensitivity. Thus adoption of a DNase I-sensitive smooth-fibre conformation is prerequisite but not in itself sufficient for the transcription of genes within that chromatin region. Despite these caveats, the term 'active chromatin' is often used to denote *all* DNase I-sensitive regions, whether or not they are actually engaged in transcription.

Ribosomal genes (Foe, 1977) transcribed by RNA polymerase I, and many protein-coding genes transcribed by polymerase II, can intercon-vert between the nucleosomal and smooth-fibre chromatin conforma-tions, depending on tissue type and developmental stage. However, this may not apply to genes transcribed by RNA polymerase III; thus the *Xenopus* oocyte-type 5S genes remain DNase I-sensitive in somatic cells where they are not transcribed (Coveney & Woodland, 1982).

2.6 Distinctive features of active chromatin

2.6.1 DNase I-hypersensitive sites

Within stretches of DNase I-sensitive chromatin, certain sites (50 to 350 bp long) appear particularly susceptible to DNase I attack (Elgin, 1982). Because these are cut in preference to other sites when very low levels of DNase I are used, it is possible to map such *hypersensitive* regions with respect to known restriction sites in the gene region under study. DNase I-hypersensitive sites are also selectively sensitive to S1 nuclease (Larsen & Weintraub, 1982).

DNase I-hypersensitive sites have been described at or close to the 5′ ends of many active genes. Examples include: (a) the 5′ ends of all five histone genes in *Drosophila* (Samal *et al.*, 1981); (b) the 5' end of the preproinsulin II gene in chromatin from a rat pancreatic insulinoma, but not in chromatin from other rat tissues (Wu & Gilbert, 1981); and (c) the 5' ends of two embryonic globin genes (the β^E and α-type U genes) in early chick embryo erythroid cells (Stalder *et al.*, 1980b; Weintraub *et al.*, 1981). In this last case, the DNase I-hypersensitive sites adjacent to the embryonic globin genes disappear when these genes are switched off later in development, while new hypersensitive sites appear in the flanking regions of the late embryonic/adult globin genes. However, these new sites are located some 2 kbp and 6 kbp upstream from the 5' end of the adult β globin gene (i.e. not immediately adjacent to it).

The presence of such sites does not always imply that the adjacent gene is being transcribed. DNase I-hypersensitive sites are always present at the 5' ends of *Drosophila* heat-shock protein genes, irrespective of whether these are being actively expressed (Keene *et al.*, 1981). This may keep the heat-shock genes in readiness for immediate activation, since their transcription can be rapidly induced by raising the ambient temperature.

There is also evidence that a pattern of DNase I-hypersensitive sites can be inherited over many cell generations under conditions where the adjacent genes are not expressed. Such sites can be induced in the globin gene region of chick fibroblasts following infection with a temperature-sensitive mutant of Rous Sarcoma Virus (RSV) at the permissive temperature of 36 °C. Under these conditions, transcription of the globin genes (normally silent in fibroblasts) is induced via the viral transforming function (Groudine & Weintraub, 1975, 1981). If the ambient temperature is shifted to 41 °C (restrictive temperature), then the viral transforming protein is inactivated and globin gene transcription ceases. However, the pattern of DNase I-hypersensitive sites in the globin gene region is propagated for at least 20 generations among such cells when grown continuously at the higher temperature. This permits rapid reactivation of globin gene transcription whenever the progeny cells are returned to the permissive temperature of 36 °C (Groudine & Weintraub, 1981). A similar mechanism could allow cells to maintain pre-selected genes in an 'available' state over many cell generations, without actually expressing them until one or more appropriate signals have been received by their progeny. Such a process may underlie the phenomenon of *determination* (or commitment) during embryonic development, a topic to be discussed in chapters 4 and 7.

A hierarchy of DNase I-hypersensitive sites has been described in the chromatin of the major vitellogenin gene (VTGII) in hen liver cells

(Burch & Weintraub, 1983). Note that VTGII gene expression is controlled by oestrogen in this tissue (see §6.4). One class of DNase I-hypersensitive sites is present within the VTGII gene and beyond its 3' end in hen liver cells which have never been exposed to oestrogen; these sites may reflect the commitment of such cells to express the VTGII gene (cf. above), since they are absent in brain or fibroblast chromatin. Three further DNase I-hypersensitive sites appear within a 0.7 kbp region next to the 5' end of the VTGII gene following oestrogen administration. Two of these are stable, and are apparently passed on from parent to daughter cells (as above), even during periods of hormone withdrawal when vitellogenin expression ceases. The third, however, is present only during oestrogen treatment, i.e. when the VTGII gene is active; this site may reflect a direct interaction with the hormone-receptor complex. Alternative sets of DNase I-hypersensitive sites have been described in the chicken lysozyme gene-region (Fritton *et al.*, 1984), a gene expressed at low levels in macrophages but at high levels in hormone-stimulated oviduct (see §6.4); again, one site unique to oviduct is apparent only during hormone stimulation.

2.6.2 Associated non-histone proteins

It is pertinent to ask what features distinguish DNase I-sensitive from relatively resistant regions of chromatin; i.e. what underlies the interconversion between smooth-fibre and beaded nucleosomal forms?

Weisbrod & Weintraub showed in 1979 that the preferential DNase I-sensitivity of the globin gene region could be abolished by eluting erythroid chromatin with 0.35 M NaCl. Among the subclass of chromatin proteins removed by this treatment, two were identified as HMGs 14 and 17, belonging to the high mobility group of non-histone proteins. When NaCl-treated erythroid chromatin was recombined with either the total eluted protein fraction or with purified HMGs 14 and 17, then DNase I-sensitivity was restored in the globin gene region. However, increased DNase I-sensitivity was not induced at non-expressed sites such as the ovalbumin gene. Nor were the globin genes rendered DNase I-sensitive when say brain chromatin was subjected to the same procedure. Thus HMGs 14 and 17 confer a DNase I-sensitive conformation on active or available chromatin regions, but do not recognise particular DNA sequences as such. Similar results have been reported for other gene sites, using nick translation (§1.7) from DNase I-generated nicks to identify 'active' chromatin regions (Gazit *et al.*, 1980).

Isolated HMGs 14 and 17 have been cross-linked onto agarose as an affinity matrix to purify 'active' chromatin fragments (Weisbrod & Weintraub, 1981). This technique permitted mapping of the HMG 14- and 17-binding sites within the active α globin gene region of chick erythroid chromatin. In general terms, those chromatin fragments which bind efficiently to the HMG 14/17 matrix are the same as those showing enhanced DNase I-sensitivity; both extend for about 1000 bp beyond the primary transcript regions of the two active α globin genes. Presumably some distinctive feature of 'active' as opposed to inactive chromatin sites is being recognised by HMGs 14 and 17. However, the same study (Weisbrod & Weintraub, 1981) also indicated that this does not involve: (a) differences in DNA:protein ratio; (b) differences in nucleosome particle size or repeat length; (c) differences in the core histone complement; (d) differences in associated non-histone proteins; or (e) markedly different levels of histone modification.

HMGs 14 and 17 appear to bind to the linker DNA, possibly replacing H1 histone (see review by Weisbrod, 1982). HMG 14 in particular undergoes cycles of phosphorylation and dephosphorylation in phase with those found for H1 (Bhorjee, 1981).

2.6.3 DNA methylation

To jump the gun at this point, one possible feature recognised by HMGs 14 and 17 as distinguishing active from inactive chromatin, is the extent of methylation in the underlying DNA. In general, the DNA sequences present in DNase I-sensitive regions of vertebrate chromatin are relatively undermethylated as compared with bulk DNA (see review by Felsenfeld & McGhee, 1982). There are complications hidden within this statement, however, since the key effects of undermethylation may be confined to a few flanking sites or may extend over a much larger region of DNA.

The predominant type of methylation in vertebrate DNA gives rise to 5-methylcytosine residues at 50–90% of CG dinucleotide sequences. However, very little of the DNA appears to be methylated in some invertebrates, including *Drosophila*. The properties of the methylase enzyme from mouse ascites cells, described by Gruenbaum *et al.* (1982), include two features of particular interest: (a) the preferred substrate of the enzyme is hemimethylated DNA (i.e. methylated on one strand but not on the other); and (b) the enzyme methylates only those C residues adjacent to G residues (i.e. CG dinucleotides). These two properties ensure that the methylation pattern is passed precisely from parent to daughter duplexes during DNA replication (fig. 23).

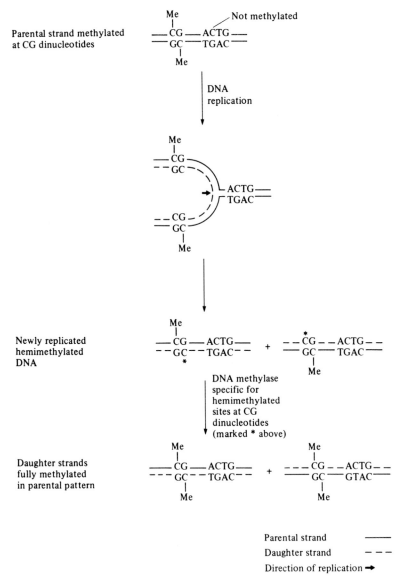

Fig. 23 DNA methylation.

Inheritance of methylation patterns, like that of DNase I-hypersensitive sites, could provide a molecular basis for determination. Stein *et al.* (1982) have shown that C-methylated or unmethylated DNA sequences of viral origin will retain their methylation pattern following integration into a host-cell chromosome, even after repeated replication through at least 100 cell generations.

The presence or absence of methyl groups on particular C residues can be demonstrated by parallel restriction digests (§1.7) using two *iso-schizomer* enzymes (from different bacteria), which recognise the same target sequence involving a CG dinucleotide. Methylation of the C residue in such a dinucleotide prevents a cut from being made by one isoschizomer enzyme, but has no effect on cutting by the other. For instance, both *Msp*I and *Hpa*II recognise the same target sequence CCGG. Cutting by *Hpa*II is prevented if the internal C residue is methylated ($CC^{Me} GG$), but this does not affect cutting by *Msp*I. By contrast, methylation of the external C residue ($C^{Me} CGG$) prevents *Msp*I but not *Hpa*II from cutting at this target site. Thus for the internal *CG* dinucleotide, *Hpa*II is C-methylation sensitive, while *Msp*I is not.

But what is the evidence that undermethylated regions in vertebrate DNA really represent active or potentially active genes? Do HMGs 14 and 17 recognise such undermethylated regions and confer upon them a DNase I-sensitive chromatin conformation? And are heavily methylated DNA regions always inactive in transcription? The answers to these questions remain unclear for the time being, but a few relevant observations are summarised below.

Gene sequences present in the 'active' DNase I-sensitive form of chromatin can be labelled by nick translation (§1.7), after limited DNase I treatment of isolated nuclei (introducing nicks only in DNase I-sensitive regions; see Levitt *et al.*, 1979). Naveh-Many & Cedar (1981) have shown that the labelled sequences generated by this method are preferentially susceptible to attack by methylation-sensitive restriction enzymes (such as *Hpa*II). Some 70% of CG dinucleotides are C-methylated in total vertebrate DNA, as compared with only 30–40% in the labelled DNA fraction representing active chromatin. Thus active gene regions are about two-fold less methylated than average.

However, it is possible that demethylation of a small DNA region (perhaps near the 5' end of a gene) might suffice to facilitate transcription, even if the rest of the gene remains heavily methylated. This is apparently the case for the *Xenopus* ribosomal gene repeats, which are highly methylated apart from two undermethylated regions in the non-tran-scribed spacer (Bird *et al.*, 1981). This situation pertains in all somatic tissues where the rDNA is actively transcribed. But in sperm cells the inactive rDNA repeats are fully methylated even at the above-mentioned spacer sites, and this may also be true for the chromosomal rDNA of oocytes. Methyl groups are lost from these two spacer sites during the first 20 hours of embryonic development, concomitant with the activation of rDNA transcription (see chapter 4). However, fully methylated sperm rDNA can be transcribed when injected into oocytes, even though the spacer sites do not become demethylated (Macleod & Bird, 1983). Thus demethylation is not prerequisite for transcription in this system, although the two processes seem to be associated *in vivo*. Two genes which encode enzymes essential for cell survival are characterised in all tissues by demethylated sites near their 5' ends, the remainder of both genes being heavily methylated (Stein *et al.*, 1983).

As with DNase I-sensitivity, the presence of undermethylated sites probably correlates with 'availability for transcription' rather than active transcription as such. Thus for example, the 5' end of the rat albumin gene is undermethylated in each of several hepatoma cell lines, some of which express albumin mRNA and protein while others do not (Ott *et al.*, 1982); however, the entire albumin gene is undermethylated and actively transcribed in normal liver cells *in vivo*. Another pertinent example is that of the chicken vitellogenin (VTGII) gene, whose expression can be switched on by oestrogen in liver but not in oviduct cells (see §6.4). Most of this gene is methylated whether or not it is being actively transcribed. Oestrogen treatment induces demethylation at a single site 0.6 kb from the 5' end of the gene, both in liver cells which express vitellogenin, and also in oviduct cells which do not (Wilks *et al.*, 1982; Burch & Weintraub, 1983). In liver, this demethylation occurs *after* the initiation of VTGII transcription and formation of the 5' DNase I-hypersensitive sites (see previous section; Burch & Weintraub, 1983). Moreover, two *Xenopus* vitellogenin genes (A1 and A2; see §6.4) are expressed in oestrogen-treated liver cells without any apparent demethylation occurring in or close to them; both remain as fully methylated in DNA from oestrogen-treated liver as in embryonic or erythrocyte DNAs (Gerber-Huber *et al.*, 1983). Thus demethylation does not seem to be prerequisite for gene expression, but may rather be consequent upon it in some systems.

A different line of evidence bearing on this question emerges from studies using the drug 5-azacytidine, which induces demethylation of C residues in the DNA. Cell cultures treated with 5-azacytidine often express genes not previously active in those cells, possibly because of demethylations affecting those genes or adjacent DNA sequences (see Felsenfeld & McGhee, 1982). However, these experiments are difficult to interpret because of the toxic side-effects of 5-azacytidine.

It is apparent from the foregoing that methylation and demethylation of C residues in CG dinucleotides does not play a universal role in regulating gene expression. Many invertebrates (e.g. *Drosophila* and *Caenorhabditis*) contain little if any methylated DNA; clearly non-expressed genes must be inactivated by other means in these organisms. Other invertebrates (such as sea urchins) contain a considerable proportion of methylated DNA sequences, but these do not seem to include any identified genes. Methylation here may permanently inactivate some parts of the genome, while expressible sequences remain unmethylated and must be subject to other regulatory mechanisms. Finally, in the vertebrates a large proportion of the DNA is methylated. The 5'-flanking sequences of many 'housekeeping' genes are always found to be demethylated, but these genes are also expressed in all tissues. This situation cannot be classed as a true regulatory system involved in on/off switching. As for vertebrate genes expressed in a tissue-specific manner, there are some cases where demethylation of 5'-flanking sequences is apparently prerequisite for gene expression, and others where it is not. The latter group is exemplified by the *Xenopus* and chicken vitellogenin genes (discussed above), where detectable demethylation does not precede active transcription. However, the globin genes provide a clear example of the former category, as demonstrated by a direct functional test. When copies of the human γ globin gene region are introduced into heterologous mouse cells, expression occurs only when the 5'-flanking sequences are demethylated and not when these sequences are fully methylated (see Busslinger *et al.*, 1983; Bird, 1984). Further examples of both categories will be needed to establish their generality and possible significance. It must be remembered on the one hand, that a correlation between demethylation and expression at a given gene site does not in itself prove a causal link. On the other hand, the presently-available range of methylation-sensitive and -insensitive isoschizomer restriction-enzymes does not permit the detection of all possible *CG* dinucleotides, since the target sites for these enzymes are larger than the dinucleotide itself. Thus while *CCGG* sequences can be readily analysed (using *Hpa* II and *Msp* I), this is not possible for certain other sequences such as *TCGT*. Methylation or demethylation at such sequences can only be detected (at present) by direct sequencing.

3

Transcription and RNA processing

Summary

The process of transcription in eucaryotes is in many respects
more complex than that in procaryotes (§3.1). Techniques of
RNA analysis discussed in section 3.2 include transcription *in
vitro*, isolation of mRNAs (by virtue of their 3′ poly (A) tag),
translation *in vitro*, reverse transcription of mRNA into cDNA
(for use as a labelled probe), and various types of RNA–DNA
hybridisation. Eucaryotic nuclei contain three distinct RNA
polymerases transcribing respectively the major ribosomal genes
(pol I), messenger RNA precursors and other heterogeneous
nuclear RNAs (pol II), and 5S and tRNAs (pol III). These
enzymes are distinguished *inter alia* by their subunit compositions
and sensitivities to the fungal toxin α-amanitin (II > III > I;
§3.3). The promoter sequences recognised by each of these
polymerases are discussed in the context of their requirements
for specific transcription *in vitro* (3.4). In general, accessory
transcription factors are necessary in addition to the purified
polymerase and a DNA template containing the appropriate
promoter sequences. These promoters are often complex sites,
involving several non-contiguous DNA sequences located in the
5′ flanking regions and/or within the gene transcribed (§3.4).

Part of each tandem repeat of the major ribosomal genes is
transcribed by pol I into a long precursor RNA, which is then
processed in several steps to yield one copy each of the 18S, 5.8S
and 28S species (§3.5). 5S gene transcripts produced by pol III
require only 3′ end-trimming to give mature 5S RNA, but tRNA
transcript processing is more complex – involving both 5′ and 3′
end-trimming, addition of the 3′ CCA terminus, extensive base
modifications, and in some cases removal of an internal *intron*

sequence by RNA splicing (§3.6). Nuclear pol II transcripts are heterogeneous in size (but generally large) and are mostly broken down rapidly within the nucleus (§3.7). Only a small percentage of these hnRNA molecules are processed to mRNA and exported to the cytoplasm. Both mRNA and hnRNA are characterised by modified 5′ ends (methylated cap structures), and in many cases by a 3′ sequence of 100–300 adenosine residues [poly(A)] added post-transcriptionally (§3.7). Cytoplasmic messengers are generally 4–5 fold smaller and considerably more stable than nuclear hnRNA chains.

Many mRNA-coding genes are split by *introns* or intervening sequences, i.e. intragenic regions of DNA not represented in the final messenger (§3.8). However, the entire gene (introns as well as the coding *exons*) is transcribed by pol II into a long hnRNA precursor, from which the intron transcripts are subsequently removed in a series of RNA splicing events. All this occurs within the nucleus, to yield a contiguous mRNA chain representing only the exons for export to the cytoplasm (§3.9). The splicing process probably involves an endonuclease/ligase complex including a short RNA molecule (U1 snRNA) as a recognition component (§3.9). Part of the U1 sequence is complementary to both sets of 'consensus' sequences found at exon–intron and intron–exon junctions in all mRNA-coding genes (§3.9). Finally, the complexity of cytoplasmic mRNA populations is compared with that of nuclear RNA populations (§3.10). Though the available evidence remains ambiguous, it seems probable that only some types of hnRNA chain act as mRNA precursors, while others are confined entirely to the nucleus.

3.1 Introduction

The complexities of genome organisation in higher organisms (chapter 1) are matched by a similar diversity of controls affecting gene expression. In general terms, the range of proteins characterising a given cell type reflects its cytoplasmic mRNA population. Selective controls over mRNA translation and degradation rates are known in several animal systems, but in most cases these serve to modulate rather than determine the basic pattern of gene expression. Messenger RNAs may also be stored in the cytoplasm prior to translation (see §§4.4, 4.5 and 4.6). It follows that many of the key events governing gene expression in eucaryotes occur within the nucleus, at one or more of the following levels:

(a) access to the gene sites (chromatin conformation);
(b) transcription itself;
(c) RNA processing;
(d) RNA export from nucleus to cytoplasm;

The first of these has already been covered (chapter 2), and little can yet be said of the last (see e.g. Nevins, 1983). The remaining two (b and c) are dealt with in the present chapter.

In attempting to understand eucaryotic gene expression, it is inevitable that guidelines should be sought among the simpler regulatory mechanisms of bacteria. Salient features of procaryotic gene control, according to the Jacob–Monod model, are as follows:

(i) arrangement of functionally related genes adjacent to one another, forming an *operon* unit;

(ii) coordinate transcription of such operons into polycistronic messenger RNAs;

(iii) control of transcription from short DNA sequences (promoter and operator) at the 5' end of each operon;

(iv) specific protein effectors (e.g. *lac* repressor) which interact with such sequences, exerting positive or negative control over the initiation of transcription;

(v) control over the site of transcriptional termination, so that different genes within an operon may be expressed at different levels (e.g. attenuation in the *trp* operon);

(vi) a single type of RNA polymerase to transcribe all types of gene.

Turning back to animal systems, the following points of contrast emerge.

(i) Identified genes are generally separated from each other by long stretches of 'spacer' DNA, even in cases where tandemly repeated genes (§§1.8, 1.9) or members of a gene family (§1.9) occur linked together on the same stretch of chromosome. True operon structures have, however, been identified in some lower eucaryotes (e.g. *Aspergillus*; Arst & Macdonald, 1975).

(ii) Because of the distances between genes, any polycistronic transcripts in eucaryotes would have to be enormously long. One case where this does occur is in the coordinate transcription of 18S + 5.8S + 28S rRNA genes to form a large polycistronic rRNA precursor (§3.5). In most other cases, the estimated sizes of primary transcript RNAs and the occurrence of promoter sites adjacent to identified genes, both argue that eucaryotic genes are transcribed individually.

(iii) Putative promoters have been identified for several genes in higher organisms (§3.4), but these are often complex sites involving DNA sequences at more than one location, some of which may lie within the gene while others lie in the upstream (5′ flanking) regions.

(iv) Hormone–receptor complexes (§§4.8, 6.3) act as positive control factors for certain steroid-regulated genes, by interacting with DNA sequence elements in their 5′ flanking regions. Another example in this category might be the TFIIIA transcription factor which binds to an internal region of the 5S gene (see §3.4).

(v) Little is yet known of the sequence requirements or accessory factors necessary for transcriptional termination in animal systems.

(vi) Animal cells have three distinct nuclear RNA polymerases designated pol I, II and III respectively, each of which transcribes a different class of genes (see §3.3). A fourth RNA polymerase, found only in mitochondria but encoded by a nuclear gene, is responsible for transcribing mDNA (§1.10).

These contrasts do not rule out a modified version of the Jacob–Monod model; they do however suggest that the regulatory networks are much more complex in higher organisms, and that a measure of control is exerted at the level of the individual gene as well as at the level of functionally related 'gene batteries'. A general model for gene regulation in higher organisms, proposed some fifteen years ago by Britten & Davidson (1969) and subsequently modified (Davidson & Britten, 1971, 1979; Davidson *et al.*, 1977), will be discussed at the end of chapter 4.

3.2 Techniques of RNA analysis

3.2.1 *In vitro* transcription

RNA synthesis *in vitro* requires a DNA template, an RNA polymerase enzyme, nucleoside triphosphate precursors (ATP, CTP, GTP and UTP), and suitable concentrations of mono- and divalent metal ions. The UTP is usually supplied in radioactive form, and the labelled RNA product is easily distinguished by its size and insolubility in trichloroacetic acid from the unpolymerised precursors. Under these conditions, both eucaryotic and bacterial RNA polymerases can initiate RNA synthesis from single-stranded nicks in the DNA template. However, the *in vitro* RNA product consists of random transcripts of all available DNA sequences, without specific initiation and termination sites as *in vivo*. Moreover, both strands of the template are transcribed with equal facility (i.e. symmetrically) *in vitro*, whereas transcription *in vivo* is usually asymmetric (from one template strand only). Several additional requirements must be met before *in vitro* transcription begins to match the *in vivo* pattern.

(a) The template must contain suitable promoter sites for the polymerase used (see §3.4). Cloned gene regions including flanking sequences are ideal for this purpose. The DNA should also be free of nicks, which can act as non-specific initiation sites for RNA synthesis.

(b) The complex RNA polymerase enzyme should be purified but fully functional.

(c) Accessory transcription factors (proteins other than the polymerase itself) are also essential in eucaryotic systems (§3.4), though only in one case has such a factor been purified and characterised in detail. In other cases, crude cell extracts meet this requirement, but the factors they contain remain ill-defined.

Specific initiation and correct asymmetry of transcription have been demonstrated *in vitro* for all three nuclear RNA polymerases, but there is less certainty about the rate of chain elongation and the site of transcriptional termination *in vitro*, particularly in the case of pol II. The use of chromatin rather than naked DNA as a template for such studies has proved fraught with problems in practice (see chapter 28 in Lewin, 1980), though pol I- and pol III-based systems are again more amenable than those involving pol II.

As an alternative to these *in vitro* systems, one can microinject cloned genes into the nuclei (germinal vesicles) of *Xenopus* oocytes, which contain large amounts of all three RNA polymerases plus RNA-processing enzymes. This technique, pioneered by Gurdon and co-workers (see e.g. Brown & Gurdon, 1977), allows both accurate transcription and subsequent RNA processing. Cloned genes can be manipulated *in vitro*, e.g. by partial deletions in the 5' flanking (upstream) regions. If such modified genes are then transcribed *in vitro* or in microinjected oocytes, the promoter activity of the deleted sequences can be gauged by measuring the extent of transcription relative to controls using the intact gene. Deletion of sequences with key promoter functions should greatly reduce or abolish transcription. This approach has allowed the mapping of eucaryotic promoter sites (see §3.4).

3.2.2 Poly(A) and messenger isolation

The first eucaryotic mRNA to be identified (Chantrenne *et al.*, 1967) was the 9S (~650 nucleotide) globin messenger, which represents more than half of the total mRNA in vertebrate erythroid cells. Separation of this messenger was initially accomplished on the basis of size; after centrifugation on a sucrose density gradient, the cytoplasmic RNA from such cells gives a discrete 9S peak distinct from the 18S rRNA and 5S rRNA/4S tRNA peaks (fig. 24). Most other cell types contain a much greater variety of messengers ranging in size from about 8S up to 20S or more, which tend to form an ill-defined smear on sucrose gradients (fig. 24). These mRNA peaks or smears are much more sensitive to mild ribonuclease treatment than are ribosomal or transfer RNAs (fig. 24), presumably reflecting a lower extent of hydrogen-bonded secondary structure.

Of the ~650 nucleotides in globin mRNA, the last hundred or so at the 3′ end consist only of adenosine residues. This poly(A) sequence is not in fact transcribed from the DNA, but rather is added on post-transcriptionally by one or more special poly(A) polymerase enzymes (see §3.7). Similar poly(A) tracts are found at the 3′ ends of most but not all eucaryotic mRNAs, though the length of poly(A) varies between different messenger species. Poly(A) is entirely absent from the ribosomal and transfer RNAs which comprise the bulk of cytoplasmic RNA in all cell types. One can therefore purify poly(A)-containing messenger RNAs by affinity chromatography on an oligo(dT)-cellulose matrix; poly(A)$^+$ RNA species will become bound to this matrix by A=T base pairing, while poly(A)$^-$ RNAs pass through (fig. 25). Poly(A)$^+$ RNAs are later released from the matrix under denaturing conditions which break H-bonds (e.g. high salt/formamide).

Several cycles of oligo(dT) chromatography are generally required in order to remove all contaminating poly(A)$^-$ RNAs from the poly(A)$^+$ messenger preparation. This technique is equally applicable to nuclear RNA, since poly(A) is added soon after transcription to some but not all nuclear transcripts (see §3.10).

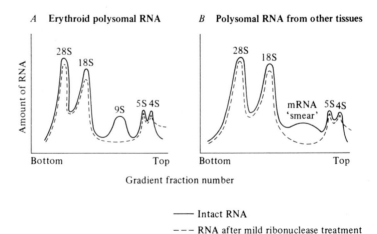

Fig. 24 Sucrose gradient fractionation of cytoplasmic RNA (from polysomes).

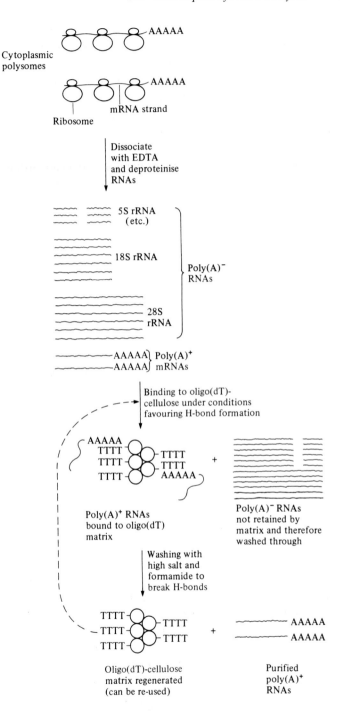

Fig. 25 Isolation of poly(A)$^+$ RNA.

Oligo(dT) chromatography purifies poly(A)$^+$ RNAs as a class, and does not distinguish between different mRNA or hnRNA species within that class. In order to isolate a specific mRNA, it is necessary either (a) to choose a tissue in which one or a small range of messengers is predominant, allowing size-separation of the desired mRNA following oligo(dT) chromatography; or else (b) to enrich the initial cytoplasmic RNA population with respect to a particular minority messenger. This latter objective can be achieved by *immunoprecipitation* of polysomes (see e.g. Schechter, 1974). Cytoplasmic groups of ribosomes engaged in translating a given type of messenger (called polysomes) will also carry nascent polypeptide chains, which can be recognised by antibodies directed against the final protein. Such antibodies should precipitate only those polysomes carrying the appropriate nascent protein chains. The immunoprecipitated polysomes are then dissociated and deproteinised to release the mRNA, tRNA and rRNA components. The desired messenger is finally purified (or at any rate enriched) by oligo(dT) chromatography as above.

Once isolated, mRNA may be identified by hybridisation (see §3.2.4), or by *in vitro* translation into protein. The latter method uses a cell-free lysate containing all the amino-acids, ribosomes, tRNAs and other factors required for protein synthesis. One can also microinject mRNAs into *Xenopus* oocytes, where they are efficiently translated for long periods (Gurdon *et al.*, 1971). The proteins produced in either system are normally labelled by including a radioactive precursor amino-acid (e.g. [^{35}S]-methionine), so that newly synthesised proteins can be separated by gel electrophoresis and the bands located by autoradiography.

3.2.3 Reverse transcription

The 'central dogma' of molecular biology, as originally formulated, proposed that the flow of genetic information is unidirectional; from DNA to RNA (transcription), and from RNA to protein (translation). However, the discovery that cells transformed by single-stranded RNA viruses (retroviruses) can transmit viral information to their progeny many generations later, suggested that a heritable DNA intermediate might be involved. This formed the basis of Temin's heretical suggestion that a viral RNA strand might be converted into a double-stranded DNA 'provirus', which could then become integrated into a host-cell chromosome and so be passed on to all that cell's descendants. Initially this idea met with a hostile reception, which was not finally dispelled until 1970, with the identification of a virus-coded enzyme (reverse transcriptase) which can copy RNA chains into complementary DNA (See Gallo, 1971). Accordingly, the first step in the central dogma must now be regarded as reversible, though the second is not.

Like other DNA polymerase enzymes, reverse transcriptase requires a preexisting 3' OH terminus from which to extend the new DNA strand in a 5'→3' direction. For *in vitro* reverse transcription of mRNA sequences, this is generally supplied as a short oligo(dT) 'primer' which can base-pair with the 3'-terminal poly(A) of the messenger (fig. 26). After synthesis of a complementary cDNA strand, the template mRNA can be destroyed by alkaline hydrolysis (which does not affect DNA).

In vivo (and also *in vitro* under certain reaction conditions) the reverse transcript can loop back on itself, allowing the enzyme to synthesise a second DNA strand complementary to the first. The end result is a duplex DNA version of the single-stranded RNA template – the 'provirus' basis of retrovirus inheritance. Double-stranded cDNAs synthesised *in vitro* provide a starting point for cloning eucaryotic messenger sequences

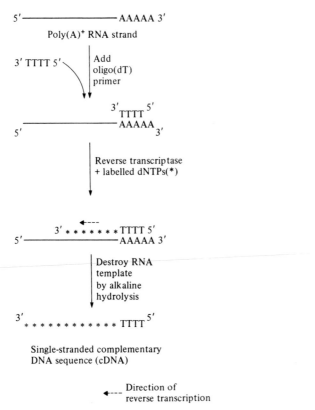

Fig. 26 Reverse transcription.

rather than genomic DNA (see §1.7). However, the single-stranded cDNA molecules resulting from one-way reverse transcription (fig. 26) have important applications of their own in molecular biology. They are complementary to the template mRNA used, and will thus base-pair specifically with that sequence during hybridisation (§3.2.4 below). Highly labelled cDNAs are therefore widely used as specific probes for the template sequence(s) in RNA or DNA populations. One drawback in early studies of this type was that reverse transcripts were often rather short owing to premature termination by the enzyme. This meant that cDNA probes were usually biassed towards the 3' terminal sequences of the template RNA. More recently, however, techniques for obtaining full-length cDNAs have been developed.

3.2.4 RNA separation and RNA–DNA hybridisation

RNA molecules, like DNA fragments, can be separated according to size by electrophoresis through agarose gels. One problem with RNA – being single-stranded – is that local regions of base-pairing can form both within and between RNA chains, the latter leading to aggregate formation. This can be overcome by size-fractionating RNA chains under denaturing conditions, e.g. in the presence of highly polar agents such as dimethyl sulphoxide or formamide, which break hydrogen bonds. The RNA equivalent of Southern blotting has been christened 'Northern' blotting, but essentially involves the same procedure (Alwine *et al.*, 1977). RNA molecules are transferred from a gel to a nitrocellulose filter or diazotised (DBM) paper sheet; this blot is then hybridised with a suitable labelled probe, and the hybridising bands identified by autoradiography.

Solution hybridisation reactions between RNA and DNA components involve denaturation and controlled annealing; it is sometimes advantageous that RNA–DNA hybrids are more stable than DNA duplexes under certain solvent conditions. The principle is basically similar to that described for DNA renaturation (see §1.3), giving R_{ot} curves where the R_{ot} value is the RNA equivalent of C_{ot}. Several types of RNA–DNA hybridisation reaction have been devised to yield different information; for a detailed discussion of the kinetics involved, see chapter 6 in Davidson (1976):

(a) *Reactions in RNA excess.* When reacting a genomic DNA tracer with a vast excess of RNA, any repetitive-sequence transcripts in the RNA component will hybridise rapidly with their corresponding template sequences in the DNA. This usually obscures the much slower hybridisation between single-copy DNA sequences and their RNA transcripts (see

McCarthy & Church, 1970). Unfortunately, the transcripts of single-copy sequences include most of the messenger RNA species, which are often the main focus of interest when comparing RNA populations. In order to study such transcripts, a tracer consisting of isolated single-copy DNA must be used, containing no repetitive sequences (see §1.5). Though repetitive-sequence transcripts are still present in the RNA excess component, they will find few if any complementary DNA strands with which to react, and thus will not obscure the hybridisation of single-copy-sequence transcripts.

Competition hybridisation can be used to compare RNA populations derived for example from different tissues. The two RNA populations are hybridised separately with a limiting amount of single-copy DNA tracer, to determine the percentage of unique DNA sequences represented in each. A mixture of both RNA populations is also hybridised to the tracer preparation, and the percentage hybridisation is again measured relative to the input of tracer. If the two RNA populations are identical, the percentage of DNA in hybrid form should be the same for both RNAs together as for each one separately (complete competition for hybridising DNA sequences). If, on the other hand, the two RNA populations are completely different, the hybridisation results for both combined should equal the *sum* of the two separate components, because each RNA population hybridises with different tracer sequences (no competition). Intermediate results indicate partial overlaps between the RNA populations compared (which is the usual outcome . . .).

A different type of RNA-excess hybridisation provides a kinetic rather than saturation analysis, i.e. in terms of RNA abundance rather than presence/absence. An entire poly(A)$^+$ RNA population is reverse-transcribed into complementary labelled cDNA strands, which are then isolated and hybridised with an excess of the original poly(A)$^+$ RNA population. Assuming that all poly(A)$^+$ RNAs were reverse transcribed with equal efficiency, the *rate* of hybrid formation between RNA and cDNA sequences should reflect the relative abundance of different RNA species in the poly(A)$^+$ population. The interpretation of such R_{ot} curves is basically similar to that for DNA reannealing (C_{ot}) curves, giving an estimate of RNA abundance and complexity within each kinetic component (see §3.10).

(b) *DNA excess reactions.* In this technique pioneered by Melli *et al.* (1971), a specific labelled RNA tracer (or a cDNA copy thereof) is hybridised with a vast excess of unlabelled genomic DNA. By comparing the hybridisation rate with known kinetic standards, it is possible to estimate the average number of copies of the tracer-coding sequence per haploid genome. This method allows one to determine copy numbers for individual genes rather than for bulk DNA classes (see §4.2).

3.3 RNA polymerases

Eucaryotic nuclei contain three distinct RNA polymerase enzymes, as first shown by Roeder & Rutter in 1969. These can be separated by ion-exchange chromatography, the order of elution from DEAE-cellulose being I, then II and finally III (sometimes termed A, B, and C respectively by European workers). The three enzymes are also distinguished from each other on the basis of: (a) their cellular location; (b) their functions *in vivo*, i.e. which genes they transcribe; (c) their sensitivity to an octapeptide toxin, α-amanitin (derived from the poisonous fungus *Amanita phalloides*); and (d) their subunit compositions. These differences are summarised in table 2 and fig. 27.

Thus nucleolar polymerase I is amanitin-resistant, and *in vivo* transcribes the major ribosomal RNAs (18S, 5.8S and 28S) encoded by the rDNA repeat unit. Nucleoplasmic polymerase II is amanitin-sensitive and synthesises all types of heterogeneous nuclear RNA and messenger RNA *in vivo*. Polymerase III, characterised by its intermediate sensitivity to amanitin and by its presence in the cytoplasm as well as nucleoplasm, transcribes small stable RNAs *in vivo*, i.e. the 5S rRNA and 4S tRNAs, and also some small viral RNAs (such as the adenovirus 5.5S VA species) in virus-infected cells.

Figure 27 compares the subunit compositions of highly purified mouse RNA polymerases I, II and III (Sklar *et al.*, 1975). Each of the multi-subunit enzymes was dissociated into its component polypeptide chains

Table 2. *Distinctions between RNA polymerases I, II and III*

Property	Polymerase I (A)	Polymerase II (B)	Polymerase III (C)
Cellular location	Nucleolus only	Nucleoplasm	Nucleoplasm and cytoplasm
Function (transcripts made *in vivo*)	18S + 5.8S + 28S rRNAs only (§3.5)	All types of mRNA and hnRNA (§3.7)	5S rRNA, 4S tRNAs and some small viral RNAs
Sensitivity to α-amanitin	Resistant (not significantly inhibited by $10\,\mu g/ml$)	Sensitive (completely inhibited by $0.1\,\mu g/ml$)	Intermediate sensitivity (inhibited by $1–10\,\mu g/ml$)

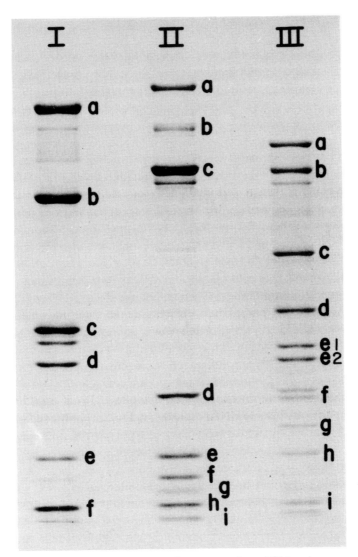

Fig. 27 Subunit compositions of mouse myeloma RNA polymerases I, II and III. Highly purified RNA polymerases I, II and III from mouse myeloma cells were dissociated into their component subunits using sodium dodecyl sulphate (SDS). These subunits were separated according to molecular size on an SDS-polyacrylamide gel, and the protein bands visualised by staining. The largest subunits are towards the top of the gel, the smaller ones lower down. Photograph kindly supplied by Prof. R. G. Roeder (Rockefeller University, New York) and reprinted with permission from the copyright holders, Academic Press. From R. G. Roeder, M. W. Golomb, J. A. Jaehning, S. Y. Ng, C. S. Parker, L. B. Schwartz, V. E. F. Sklar & R. Weinmann (1977) Animal Nuclear RNA Polymerases. In *Receptor and Hormone Action* vol. 1 (ed. B. W. O'Malley & L. Birnbaumer), pp. 195–236, Academic Press, New York.

by detergent treatment (sodium dodecyl sulphate; SDS); these subunits were then separated according to molecular size on an SDS-polyacrylamide gel, giving the patterns of stained bands shown in fig. 27.

The larger subunits of these enzymes are all different in size; thus each is probably specific to only one of the three polymerases. However, some of the small subunits are very similar in size, and may perhaps be common to two or even all three enzymes. The different band intensities in fig. 27 may be due in part to certain subunits occurring in two or more copies per enzyme complex; very faint bands may represent contaminants, or possibly subunits that dissociate easily from the main complex and are thus partly lost during purification. The functions of the various eucaryotic polymerase subunits have yet to be defined, although mutations have been identified in *Drosophila* (Greenleaf *et al.*, 1979, 1980) and in cultured mammalian cells (Ingles, 1978) which confer an amanitin-resistant phenotype by altering one subunit of polymerase II.

Amanitin clearly distinguishes between the three eucaryotic RNA polymerases *in vitro* and in isolated nuclei, but administration of the toxin *in vivo* additionally causes an indirect inhibition of polymerase I activity (Tata *et al.*, 1972). During the isolation of chromatin or nuclei, most of the RNA polymerase molecules engaged in transcription will become arrested part-way through synthesising an RNA chain. These polymerases will be reactivated under *in vitro* transcription conditions (§3.2.1), so completing already-initiated RNA chains. This *endogenous* polymerase activity must be clearly distinguished from activity due to added (exogenous) RNA polymerases when analysing the RNA products from *in vitro* systems using chromatin or isolated nuclei.

The functions of the three RNA polymerases *in vivo* and the processing of their transcripts will be considered below in sections 3.5–3.10. But since each polymerase transcribes a different class of gene *in vivo* (table 2), we will begin with a brief account of the respective promoter sites recognised by these enzymes (§3.4).

3.4 Eucaryotic promoters

Functional eucaryotic promoter sites have been delimited by means of transcription assays following partial deletion of coding and/or flanking regions from an appropriate cloned gene (§3.2.1). A 'promoter null' mutation is one from which little if any transcription is initiated, implying that some part of the deleted region has promoter function. A whole series of overlapping deletions permits the promoter site(s) to be mapped in detail, although the sequences thereby identified *in vitro* may not represent the full promoter region required *in vivo*. It will be convenient to consider the *in vitro* specificity of transcription by each RNA polymerase in conjunction with its promoter requirements, since the two are to some extent interdependent.

3.4.1 Polymerase I promoters

The location of RNA polymerase I in the nucleolus, together with the fact that rRNA synthesis in isolated nuclei is amanitin-resistant (Reeder & Roeder, 1972), leads to the conclusion that pol I transcribes the nucleolar rDNA repeats *in vivo*. Isolated nucleoli *in vitro* continue to transcribe and process rRNA precursors much as *in vivo* (Grummt & Lindigkeit, 1973). The availability of DNA enriched in ribosomal genes (isolated as a $G{\equiv}C$-rich density satellite) allowed the following question to be posed: can purified pol I recognise its correct promoter sites among the rDNA repeat units and initiate specific rRNA transcription from them *in vitro*? Early attempts to answer this question (e.g. Beebee & Butterworth, 1974, in *Xenopus*) met with only partial success. In a yeast-based system, Cramer *et al.* (1974) showed that transcription of the correct rDNA strand *in vitro* was only two-fold greater than transcription of the incorrect strand. More recently, cloned mouse rDNA has been transcribed specifically and asymmetrically by RNA polymerase I *in vitro*, but only in the presence of a crude extract from rapidly growing cells (Grummt, 1981). When several similar systems from different organisms were compared, it was found that the 'accessory factors' present in these crude cell extracts promote accurate pol I transcription only in homologous systems, i.e. these factors are species-specific (Grummt *et al.*, 1982).

Detailed restriction mapping of the non-transcribed (NTS) and external transcribed (ETS) spacer sequences in a cloned *Xenopus laevis* rDNA unit (Boseley *et al.*, 1979) has identified three putative promoter regions, two in the NTS and one at the NTS–ETS border. Moss (1982) has since extended this analysis by partial deletion mapping and microinjection of *X. laevis* rDNA into *X. borealis* oocytes; this study showed that the functional promoter at the NTS–ETS border extends from -145 bp to $+16$ bp relative to the initiation site used *in vivo* (position $+1$). Of this region, only 13 bp (-7 to $+6$) are essential for transcription after microinjection of rDNA into *Xenopus* oocyte nuclei, while a much larger site (-142 to $+6$) is necessary for transcription of rDNA *in vitro* by nuclear homogenates. This has led to the suggestion of 'nested' control regions in the rDNA promoter (Sollner-Webb *et al.*, 1983). Moreover, when cloned rDNA is injected into *Xenopus* fertilised eggs, transcription is activated 5–10 fold more efficiently if spacer sequences lying over 1150 bp upstream from the initiation site are included (Busby & Reeder, 1983); these sequences may have 'enhancer-like' properties (Labhart & Reeder, 1984). Promoter analysis of the *Drosophila* rDNA unit has shown that sequences with promoter activity lie between positions −43 and −27 bp, and also between −18 and +20 bp relative to the initiation site (Kohorn & Rae, 1983a). The first four base pairs of the ETS (+1 to +4) are of major importance for promoter function (Kohorn & Rae, 1983b).

3.4.2 *Polymerase II promoters*

Endogenous transcription of specific messenger RNA sequences continues in isolated nuclei, but is inhibited by low levels of α-amanitin, confirming that polymerase II transcribes mRNA precursors. Examples include histone mRNAs in sea urchin embryo nuclei (Shutt & Kedes, 1974), and ovalbumin mRNA in hen oviduct nuclei (Nguyen-Huu *et al.*, 1978). There have been many claims that active mRNA-coding genes in isolated chromatin can be transcribed specifically by added pol II or even bacterial RNA polymerase, yielding *in vitro* RNA products similar to those synthesised *in vivo*. However, these experiments are almost impossible to interpret. Endogenous RNA polymerase activity (completing already-initiated RNA chains), as well as completed transcripts still associated with the chromatin, will both contribute *in vivo*-like RNA species. The use of mercury-substituted RNA precursors during chromatin transcription studies enables *in vitro*-synthesised RNAs (Hg-labelled, and therefore retained by a sulphydryl-agarose matrix) to be distinguished from preexisting endogenous RNAs (unlabelled, hence not retained). However, it has since transpired that *E. coli* RNA polymerase can copy endogenous RNA chains *in vitro* into complementary RNA strands; these will be Hg-labelled and therefore retained, along with the template RNA chains to which they remain hydrogen-bonded as RNA duplexes (Zasloff & Felsenfeld, 1977). When such artefacts are taken into account, the proportion of specific RNA transcripts which can be attributed to exogenous polymerase activity *in vitro* is usually much lower than the proportion of that same sequence in nuclear RNA *in vivo*. Probably the exogenous polymerases merely transcribe smooth-fibre regions of chromatin at random (giving a low level of 'specific' product), whether or not those regions are extensively transcribed *in vivo*.

A clearer definition of the factors required for specific transcription by polymerase II became possible with the advent of cloned DNA sequences containing mRNA-coding genes plus their 5′-flanking (upstream) sequences. *In vitro*, purified polymerase II does not transcribe specifically from the promoter sites in such DNAs unless accessory transcription factors are also supplied. Such factors are present in several types of crude cell extract, which contain neither DNA nor RNA polymerase II activity of their own. The first success along these lines was obtained with the adenovirus 2 genome, whose late genes (those expressed in the later stages of viral infection) are transcribed *in vivo* by host-cell pol II using a single strong promoter site. In the presence of accessory factors, purified pol II can recognise and transcribe from this promoter in naked adenovirus 2 DNA *in vitro* (Weil *et al.*, 1979). Specific initiation of transcription by pol II has since been reported for several mRNA-coding genes, including the mouse β globin gene (Luse & Roeder, 1980) and the chicken ovalbumin and conalbumin genes (Wasylyk *et al.*, 1980). In all cases a crude cell extract is essential for asymmetric transcription and accurate initiation at the site used *in vivo*. Such extracts may contain factors acting only on particular kinds of gene; e.g. a HeLa-cell factor enhancing transcription from SV40 viral promoters (Dynan & Tijan, 1983), and a transcription factor specific for the *Drosophila* heat-shock genes which binds to their upstream regulatory regions (Wu, 1984).

Deletion mapping of the promoter sites for mRNA-coding genes has mostly relied on: (i) *in vitro* pol II-transcription systems, e.g. for the rabbit β globin promoter (Grosveld *et al.*, 1981a); or (ii) microinjection into *Xenopus* oocyte nuclei, e.g. for a sea urchin histone H2A promoter (Grosschedl & Birnstiel, 1980a, b). It has also proved possible to introduce cloned genes into mammalian cells by means of *cotransformation*. Mouse cell lines lacking thymidine kinase (TK) activity can be 'transformed' through their ability to take up DNA containing a Herpes virus TK gene. The resulting TK^+ transformant cells can be selected from among their untransformed TK^- colleagues under culture conditions where TK activity is essential for cell survival. In cotransformation, a foreign gene is first covalently linked to one end of the viral TK gene; this is followed by uptake of the 'hybrid' DNA into TK^- cells and selection of the TK^+ transformants as above. Dierks *et al.* (1981) prepared a series of cloned rabbit β globin genes containing varying lengths of 5' flanking sequence; these were then linked to Herpes TK genes for subsequent cotransformation. In the selected TK^+ cells, the extent of β globin mRNA expression depended on the length of 5' flanking sequence retained in that clone, allowing *in vivo* promoter-mapping for the β globin gene (Dierks *et al.*, 1983).

Results from these different mapping approaches are not entirely consistent, but at least some common features emerge from the pol II promoter sequences studied to date:

(a) The Goldberg-Hogness or TATA 'box' sequence (actually ranging from ATA to TATAAATA) is located some 20–30 bp upstream from the transcription initiation site of most mRNA-coding genes. This sequence does not seem absolutely essential for promoter activity except in some fully *in vitro* systems (Corden *et al.*, 1980). However, in its absence a variety of abnormal 5′ termini are found among the transcripts from a cloned sea urchin H2A gene (Grosschedl & Birnstiel, 1980a). Thus the TATA box may act as a locator element for RNA polymerase II, directing initiation at a nearby downstream site.

(b) The (C)CAAT 'box' is located between positions −80 and −70 in the 5′-flanking region upstream from the initiation site (+1) of many pol II-transcribed genes. The importance of this site for promoter activity again varies somewhat according to the mapping technique used (compare Grosveld *et al.*, 1981a and Dierks *et al.*, 1983).

(c) Sequences further upstream from the initiation site may also influence transcription. In the case of the rabbit β globin gene, such a region occurs around position −100; this comprises two closely related sequences of 14 and 15 bp (imperfect tandem repeat), both of which are necessary for full promoter activity (Dierks *et al.*, 1983). In the case of the sea urchin H2A gene, a spacer DNA sequence located between positions −111 and −446 bp (segment E) dramatically influences the rate of H2A transcription, but it can function efficiently even when reversed in orientation (see Grosschedl & Birnstiel, 1980b). Part of this sequence is related to the long terminal repeats (LTRs) which enhance the transcription of many animal virus genomes (e.g. SV40). LTR-like sequences acting as transcriptional enhancers affect sequences extending for several kbp both upstream and downstream, hence they are not necessarily confined to the 5′-flanking region of the gene. In the case of the heavy-chain immunoglobulin gene, an LTR-like sequence has been identified at an internal intron site (see §3.8) separating the J_H and C_H regions of the gene as rearranged in B lymphocytes (see §4.2; Gillies *et al.*, 1983, and Banerji *et al.*, 1983). This particular enhancer sequence is only effective in lymphoid cells, i.e. is tissue-specific. The possibility that tissue-specific transcriptional enhancer elements may exist within or even at some distance from genes expressed at high levels in particular cell types, may well prove crucial to our understanding of the differentiation process (see §4.8).

3.4.3 Polymerase III promoters

The synthesis of low molecular weight stable RNAs (5S and pre-4S) in isolated nuclei or chromatin is mediated by endogenous RNA polymerase

III activity, sensitive to moderate but not low levels of α-amanitin (Marzluff *et al.*, 1974; Marzluff & Huang, 1975). Very high levels of 5S and tRNAs are expressed in young *Xenopus* oocytes (see §4.4), hence this has become the system of choice for studying polymerase III function. Specific and asymmetric transcription of 5S RNA from *Xenopus* oocyte chromatin was obtained with exogenous pol III, but not with other types of added RNA polymerase (Parker & Roeder, 1977). However, symmetrical non-specific transcription resulted when cloned 5S DNA was used as a template for purified pol III *in vitro* (Parker *et al.*, 1977). The specificity of 5S RNA transcription *in vitro* can be greatly improved by including a post-chromatin supernatant fraction from immature oocytes, together with the purified pol III and cloned 5S DNA (Ng *et al.*, 1979). As with the pol I and II systems, this crude fraction is itself devoid of DNA or pol III activity, but includes 'accessory transcription factors'. Specific *in vitro* transcription by pol III in the presence of such factors had previously been demonstrated by Wu (1978), using naked adenovirus 2 DNA as template. The 5.5S VA genes of this viral genome are transcribed selectively by pol III *in vivo*, and also *in vitro* when a crude cell extract is provided.

The role of accessory transcription factors is more clearly defined in the pol III/5S system than in the various pol I and pol II systems discussed above. TFIIIA is the principal (but not the only) factor required for accurate transcription of oocyte-type 5S genes by pol III; it is also an abundant protein in young *Xenopus* oocytes. In brief, this same TFIIIA protein not only binds to the 5S gene itself (so facilitating its accurate transcription by pol III), but also binds to the 5S rRNA transcripts, forming 7S ribonucleoprotein complexes which are an abundant storage product in the cytoplasm of immature oocytes (Honda & Roeder, 1980; Pelham & Brown, 1980). A related (or identical?) transcription factor plays much the same role in *Xenopus* somatic cells, where it acts on the somatic-type 5S genes and their RNA products (Pelham *et al.*, 1981). A large store of TFIIIA protein in young oocytes permits active transcription of the 5S genes and storage of their products, but as the TFIIIA store becomes depleted so the rate of 5S synthesis declines (§4.4).

The TFIIIA factor binds to an *internal* region of the oocyte-type 5S gene, between positions +45 and +96 in the 120 bp sequence (Engelke *et al.*, 1980). Possibly two molecules of TFIIIA bind cooperatively per 5S gene (Hanas *et al.*, 1983). Binding at this site is independent of pol III, suggesting a role for TFIIIA as a locator directing accurate initiation by pol III at an upstream site. The same conclusion is reached by promoter mapping experiments, using partial deletions of a cloned 5S gene and *in vitro* transcription in an oocyte extract system. The 5'-flanking regions and first 50 base pairs of the gene itself can be deleted without abolishing the synthesis of a 5S-sized RNA product (Sakonju *et al.*, 1980). Apparently pol III can initiate at any sequence located 50 bp upstream from the boundary at position +50. Similarly, the 3'-flanking regions and last 37 bp of the 5S gene do not affect correct initiation by pol III (Bogenhagen *et al.*, 1980). Thus the region showing pol III promoter activity in fact lies *within* the 120 bp 5S gene, extending from position +50 to +83. This is the same region as that which binds TFIIIA *in vitro* (Engelke *et al.*, 1980). Termination of 5S RNA transcription occurs at a consensus sequence including a small cluster of T residues; this sequence is recognised by pol III itself in the absence of accessory factors (Cozzarelli *et al.*, 1983).

Polymerase III also transcribes tRNA genes *in vivo* and *in vitro*, though TFIIIA does not interact with these genes nor affect their transcription. Promoter mapping studies suggest a split promoter site lying within the tRNA gene. For the *Xenopus laevis* rRNA$_1^{Met}$ gene, the identified promoter regions span positions +8 to +13 (box A) and +51 to +72 (box B) respectively (Hofstetter *et al.*, 1981); the corresponding sites

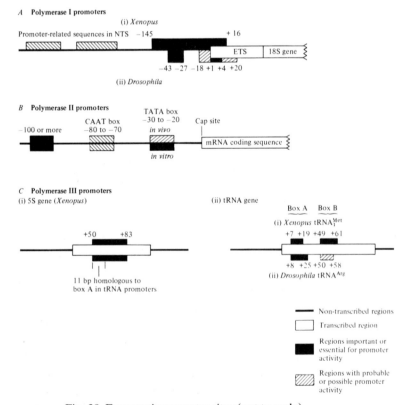

Fig. 28 Eucaryotic promoter sites (not to scale).

in the tRNA$^{Leu}_{CUG}$ gene of the same species occupy positions +13 to +20 and +51 to +64 (Galli *et al.*, 1981). These sequences are well conserved and occupy homologous positions in all eucaryotic tRNA genes analysed to date. The distance between these two promoter elements appears crucial for promoter activity, since deletions between positions +14 and +50 in the tRNA$^{Met}_1$ gene abolish transcription, although insertions of extra DNA between these same positions do not (Hofstetter *et al.*, 1981). Specific mutagenesis (changing C≡G to A=T base pairs) within this same tRNA$^{Met}_1$ gene confirms that the sequences from +7 to +19 and from +49 to +61 are the major determinants of promoter activity (Folk & Hofstetter, 1983). In the *Drosophila* tRNA^{Arg} gene, the sequence from positions +8 to +25 is essential for promoter activity, but a second region from +50 to +58 is also necessary for full transcriptional efficiency (Sharp *et al.*, 1981).

Detailed mapping of the 34 bp internal promoter of the 5S gene shows that it too is split into two elements, the first 11 bp being structurally and functionally homologous to the box A component of tRNA promoter sites (Ciliberto *et al.*, 1983). A summary of recent findings on eucaryotic promoters is given in fig. 28.

3.5 Synthesis and processing of polymerase I products

Ribosomal RNA synthesis and processing within the nucleolus provides a relatively simple model system for *in vivo* and *in vitro* study, since the rRNA products are abundant, the rDNA is tandemly repeated (§1.8), a single RNA polymerase (pol I) is involved, and nucleoli can be readily isolated from lysed nuclei.

When isolated nucleoli are incubated *in vitro* under optimal conditions for pol I transcription, a series of labelled RNA products is obtained, corresponding in size to those found *in vivo* in the nucleolus (Grummt & Lindigkeit, 1973). The largest major product sediments at 45S (approx 12000 nucleotides) in mammalian systems, and represents the entire transcription unit, including the external and internal transcribed spacers as well as the 18S, 5.8S and 28S coding regions, but excluding the non-transcribed spacer (see fig. 15). In other animals the sequence arrangement of this large precursor is similar, but the overall length may be considerably shorter. These differences are due in part to a lower proportion of transcribed spacer. Thus in mammals only 52% of the 45S (12000 base) precursor is conserved as 18S + 5.8S + 28S rRNAs, whereas in amphibians around 80% of the 40S (8000 base) precursor is conserved. In *Drosophila* and sea urchins the rRNA primary transcript is even smaller – around 34S.

A **Pathway of rRNA processing in mouse L cells (after Lewin 1980)**

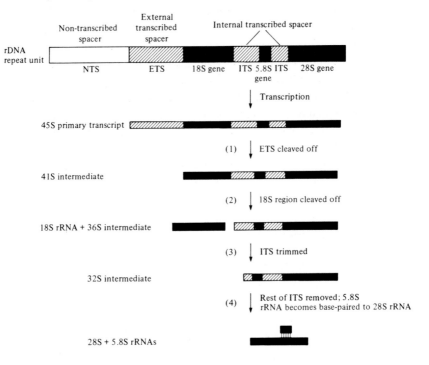

Fig. 29*A* Pathway of RNA processing in mouse L cells (after Lewin, 1980).

B **Pathways of rRNA processing in other systems**

The same sequence of processing events is followed in *Xenopus* and *Drosophila*, but the sizes of the precursor and intermediates are smaller, viz.:

Mouse (above) 45S $\xrightarrow{(1)}$ 41S $\xrightarrow{(2)}$ 18S + 36S $\xrightarrow{(3)}$ 32S $\xrightarrow{(4)}$ 28S

Xenopus   40S $\xrightarrow{(1)}$ 38S $\xrightarrow{(2)}$ 18S + 34S $\xrightarrow{(3)}$ 30S $\xrightarrow{(4)}$ 28S

Drosophila   34S $\xrightarrow{(1)}$ 33S $\xrightarrow{(2)}$ 18S + 28S $\xrightarrow{(3)}$ 27S $\xrightarrow{(4)}$ 26S

In Hela cells of human tumour origin, the order of cleavages (2) and (3) is reversed, giving the following pathway:

45S ⟶ 41S ⟶ 20S + 32S
                    ↓   ↓
                  18S   28S + 5.8S

Fig. 29*B* Pathways of rRNA processing in other systems (after Lewin, 1980).

The large primary transcript is cleaved in several stages during rRNA processing to produce 18S, 28S and 5.8S rRNAs (this last is found base-paired to the 28S rRNA within ribosomes). The intermediates belong to several discrete size classes, e.g. 41S, 36S and 32S in mammalian systems (see fig. 29A), and the sequence of processing events has been fully worked out in several animal species (summarised in fig. 29B).

About 110 methyl groups are added post-transcriptionally to each 45S precursor molecule; these seem to be located mainly within the 18S and 28S coding regions, since most of them are conserved during subsequent processing. It seems probable that methylated regions of the primary transcript are left intact by the processing enzymes, while non-methylated regions (representing the ETS and ITS) are degraded.

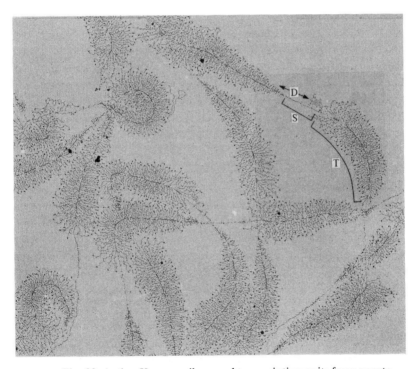

Fig. 30 Active *Xenopus* ribosomal transcription units from oocyte nucleoli, after spreading for electron microscopy. Each DNA axis (D) includes both transcribed regions (T) in the process of synthesising 40S ribosomal precursor RNA (seen as side branches of increasing length) and silent spacer (S) regions, alternating in a regular tandem array. Magnification ×15 100. Electron micrograph kindly supplied by Prof. O. L. Miller Jr. (University of Virginia, Charlottesville) and reprinted with permission from the copyright holders, A. R. Liss Inc. From O. L. Miller Jr. & B. R. Beatty (1969) *J. Cell. Physiol.* **74** suppl. 1, 225–32.

The extrachromosomal nucleoli of *Xenopus* oocytes have provided particularly favourable material for EM spreading studies of actively transcribed gene regions (Miller & Beatty, 1969). Characteristically, the rDNA axis is extended (not packaged into a beaded string of nucleosomes) and alternates between transcribed and non-transcribed regions. Where active transcription is occurring, the rRNA primary transcripts (40S) are seen to form side-branches of increasing length, originating from RNA polymerase I molecules crowded together along the rDNA axis within each transcription unit. The resultant 'Christmas tree' patterns are shown in fig. 30. The chromatin structure of insect ribosomal genes has been compared at different developmental stages when rRNA synthesis is respectively active or repressed. Electron microscopy reveals that the rDNA is mostly packaged into nucleosomes in the latter case, but forms 'Christmas tree' structures in the former (Foe, 1977); intermediate structures without nucleosomes or side-branches are also seen, perhaps equivalent to smooth-fibre non-transcribed chromatin.

3.6 Synthesis and processing of polymerase III products

Of 5S rRNA little more need be said. It is transcribed initially as a slightly longer precursor species containing about eight extra nucleotides at the 3' end. This precursor is extremely short-lived *in vivo*, presumably as a result of rapid end-trimming to generate mature 5S rRNA. However, the steady-state concentration of the precursor can be increased under conditions which inhibit processing, either *in vivo* (e.g. during heat shock; Rubin & Hogness, 1975) or *in vitro* (e.g. in isolated nuclei treated with ribonuclease inhibitor; Hamada *et al.*, 1979).

The tRNAs are also transcribed in the form of slightly longer precursors (pre-4S or 4.5S RNA). The extra sequences present in pre-tRNA transcripts are sometimes confined to the 3' and 5' termini, and could be removed by end-trimming as in the case of 5S RNA. However, many tRNA precursors also lose an *internal* sequence during processing, the two RNA fragments on either side being rejoined (spliced) to create a mature tRNA molecule. The genes encoding tRNAs of this type are said to be split or *interrupted*; i.e. they contain an *intron* sequence not represented in the mature tRNA, separating two *exon* sequences which are represented in the final product. As we shall see (§3.8 below), such interruptions are common in mRNA-coding genes. Only some types of tRNA gene possess introns; where they do occur they are invariably short (10 to 30 bp), and are located adjacent to the anticodon loop sequence. Such tRNA introns are not flanked by the short 'consensus sequences' present at all exon–intron and intron–exon junctions in split mRNA-coding genes (§3.9). This contrast suggests that a different excision/splicing mechanism is used for removing short intron transcripts from tRNA precursors. Probably this process is directed by features of the adjacent exons and/or the tertiary structure of the tRNA precursor molecule (see Cech, 1983).

In addition to splicing and 5'/3' end-trimming, a considerable number of bases in the tRNA molecule become modified e.g. by methylation, and the characteristic 3'-CCA terminus is added post-transcriptionally in eucaryotes. The order of events in tRNA processing has been elucidated by microinjecting cloned tRNA genes into *Xenopus* oocyte nuclei. For instance, the yeast tRNA^{Tyr} gene is transcribed into precursor tRNA^{Tyr}, which is then processed via 5' end-trimming, 3' end-trimming, addition of the CCA 3'-terminus, and finally intron removal/exon splicing (in that order). Several base modifications are also made in a strict temporal order which correlates precisely with the changes in precursor size during processing (Melton *et al.*, 1980). All of these events occur within the nucleus, so that only mature tRNA^{Tyr} molecules are exported to the cytoplasm. Presumably neither the processing enzymes nor the sequences they recognise are species-specific, since a yeast tRNA gene is correctly transcribed and its product processed in *Xenopus* nuclei.

3.7 hnRNA and mRNA: polymerase II products

3.7.1 hnRNA

RNA polymerases I and III between them account for the synthesis of all ribosomal and transfer RNAs and their precursors. This leaves one polymerase (II) to transcribe all other types of RNA, including the cytoplasmic messenger RNAs. Within the nucleus, amanitin-sensitive polymerase II transcribes a heterodisperse array of RNA species ranging up to sizes considerably larger than the 45S rRNA precursor. Radioactively labelled RNA precursors are rapidly incorporated into this *heterogeneous nuclear (hn)* RNA during the first 5 to 15 minutes of labelling, before significant amounts of label enter the nucleolar rRNA precursor. However, this hnRNA is mostly unstable; much of it is rapidly broken down within the nucleus, as shown by pulse-chase labelling experiments (see fig. 31), where the radioactive precursor (pulse) is washed out with an excess of the same precursor in unlabelled form (chase). Estimates of average hnRNA half-life are mostly in the range 10 to 30 minutes (Brandhorst & McConkey, 1974), though a proportion may be degraded much more rapidly. Only a small percentage of the label incorporated into hnRNA can be chased into cytoplasmic mRNA, while over 90% is turned over within the nucleus. In erythroid cells, for instance, only 1.7% of the total hnRNA is exported into the cytoplasm, of which about half (0.8%) is globin messenger RNA (Tobin *et al.*, 1978). Two reasons probably account for this massive discrepancy: (a) only some hnRNA species are actual precursors of mRNA (see §3.10); and (b) only a part of each such precursor is conserved for export to the cytoplasm in messenger form (see §3.9).

The size of hnRNA primary transcripts has been a source of some controversy. Since turnover is rapid, only pulse-labelled hnRNA extracted from nuclei under conditions which inhibit RNA degradation, can provide a reliable estimate of size. Even here some breakdown may already have occurred, leading to an underestimate of primary transcript size. On the other hand, long chains of single-stranded RNA will contain some complementary sequences, resulting in both intrastrand and interstrand base-pairing. The latter could link two or more separate chains into an H-bonded aggregate which will sediment faster (i.e. appear larger) than any of its component molecules, so leading to an overestimate of transcript size. H-bonded aggregates can be dissociated into single chains under denaturing conditions, usually in the presence of high concentrations of polar agents such as formamide or dimethyl sulphoxide (DMSO). These treatments indeed reduce the maximum size of hnRNA, though a proportion of chains longer than 45S is still detectable (see fig. 31). The average denatured chain length for hnRNA is 8–10000 nucleotides, some 4–5 fold larger than typical mRNA molecules.

A more detailed consideration of the precursor-product relationship between some hnRNAs and messenger RNAs will follow in sections 3.9 and 3.10.

Within the nucleus, hnRNA chains acquire 5′-terminal methylated cap structures, and some 25–30% also become polyadenylated at their 3′ termini. Both structures are characteristic of cytoplasmic mRNAs, and are discussed in more detail under that heading; their presence on a significant proportion of hnRNA molecules suggests a precursor–product relationship in which the 5′ and 3′ termini of hnRNA are both conserved in mRNA. However, 5′ caps or 3′ poly(A) might also be added to the new termini created by internal cleavage of longer nuclear RNA chains, in which case the presence of these structures at the ends of both mRNA and hnRNA molecules could be merely fortuitous.

HnRNA molecules often contain a sequence of oligo(U) located close to the 5′ terminus (Molloy *et al.*, 1974). This sequence comprises about 20

A **Size of pulse-labelled hnRNA under denaturing and non-denaturing conditions**

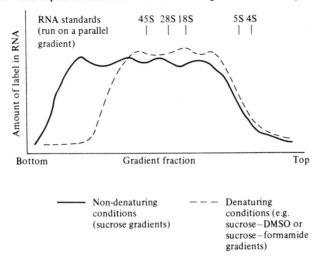

B **Breakdown of hnRNA (non-denaturing conditions)**

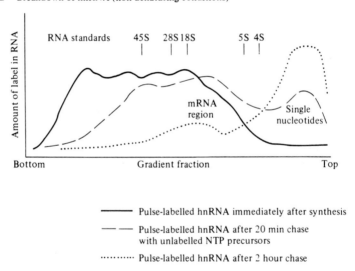

Fig. 31 hnRNA (diagrammatic).

successive uridine residues, and is transcribed from the DNA rather than added post-transcriptionally. It apparently becomes base-paired with the 3'-terminal poly(A), and this duplex region may perhaps provide a binding site for RNA processing enzymes and other macromolecules. Most hnRNA chains include repetitive as well as unique sequences, interspersed in the pattern characteristic of the parent genome (Smith *et al.*, 1974). Inverted repeat sequences also occur, leading to intramolecular duplex stems ('foldback' RNA) and simple or complex loop structures.

3.7.2 *mRNA*

The general principles of mRNA isolation have been described above (§3.2.2). Several characteristic features of cytoplasmic messengers in animal systems are summarised in this section.

Methylated cap structures are found at the 5' termini of all mRNA and hnRNA molecules *in vivo*. Thus the 5' purine triphosphate residue (ATP or GTP) which initiates each primary transcript must subsequently become modified by capping. This probably occurs during transcription of the rest of the RNA chain, since even nascent hnRNA chains contain capped 5' termini (Babich *et al.*, 1980). In essence, the capping process protects the 5' terminus of an RNA chain by linking a guanosine residue in reverse orientation onto the initial purine residue of the original transcript. The reversed G residue is then methylated in at least one position, usually to give 7-methylguanosine (see fig. 32).

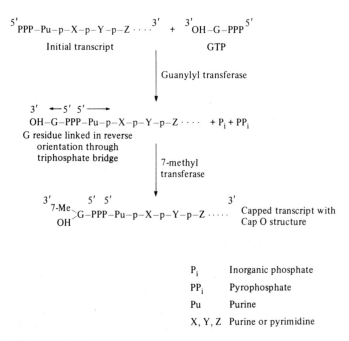

Fig. 32 Capping.

This cap 0 structure is characteristic of all yeast mRNAs and a minority of *Drosophila* mRNAs. In most animal cells, however, the 5'-terminal nucleotides become methylated at one or more further sites, usually in the 2' 0-ribose position. In cap 1 structures, only the purine residue which initiated transcription is thus modified, whereas in cap 2 structures the first two residues of the original transcript become methylated. Type 1 cap structures predominate in animal cells, and are mostly in the form ^{7Me}G-ppp-Pu^{Me}-pX-pY ... In a minority of RNA chains initiated with A, this residue becomes dimethylated, the second methyl group being attached through the N6 position. Other A residues at various internal sites in the mRNA chain may also become methylated at this position. Some capped mRNAs are found with methylated pyrimidine residues adjacent to the 7-methylguanosine added post-transcriptionally. Since transcription always initiates with a purine (A or G) residue, this suggests that cap structures may also be formed on some 5' termini resulting from internal chain cleavage. Thus a capped structure is not necessarily equivalent to the 5' end of a primary transcript. Capping not only protects the RNA chain from 5' degradation, but also plays a role in ribosome binding (see chapter 23 in Lewin, 1980). Messenger RNAs from which the caps have been removed cannot be translated efficiently into protein.

Polyadenylation is also a post-transcriptional process, whereby successive adenosine residues are added onto the 3' terminus of an RNA chain to generate a poly(A) 'tail' of 100 to 300 bases. This 3' terminus is probably already processed to some extent, since poly(A) addition usually occurs at sites upstream from the termination point of transcription (see Nevins, 1983). Characteristically, poly(A) is located at a site some 10–30 bases downstream from the sequence AAUAAA in the mRNA chain; this sequence probably performs a signalling role for poly(A) addition (Nevins, 1983). One or more nuclear poly(A) polymerase enzymes are responsible for polyadenylation, differing as regards inhibitor sensitivity from the major RNA polymerases. Thus cordycepin (3' deoxyadenosine) inhibits poly(A) polymerase, but does not affect RNA synthesis mediated by pol II, whereas α amanitin has the opposite effect. As mentioned earlier (§3.2.2), poly(A) is absent from all pol I and pol III products *in vivo*, but is characteristically present on some hnRNAs and most mRNAs; exceptions include the poly(A)$^-$ histone mRNAs (Adesnik & Darnell, 1972), and the poly(A)$^-$ messenger encoding a major 45000-dalton protein in sarcoma ascites cells (Sonnenschein *et al.*, 1976). Although the major histone mRNAs are normally poly(A)$^-$, there is some evidence that they are polyadenylated during oogenesis in *Xenopus*. Furthermore, the messenger for H5 histone (found in nucleated erythrocytes; see §2.2) is poly(A)$^+$. In both situations, histone mRNA synthesis occurs in the absence of DNA replication, whereas in dividing cells these two processes are coupled and the histone mRNAs are poly(A)$^-$ (see Molgaard *et al.*, 1980).

The exact functions of poly(A) remain unclear. It may protect the mRNA against 3′ degradation; thus globin mRNA from which the poly(A) has been removed is less stable than poly(A)$^+$ globin mRNA, following injection into *Xenopus* eggs (Huez *et al.*, 1974). However, some normally poly(A)$^-$ messengers appear relatively stable in the cytoplasm. Poly(A) may also serve to distinguish mRNA precursors from other types of hnRNA during processing within the nucleus, or may even act as a kind of 'export licence' for transferring processed messengers to the cytoplasm. In support of these ideas, Darnell and co-workers have shown that the 25% of hnRNA chains which become polyadenylated in the nucleus include some 70% of the specific mRNA sequences identified using cloned probes (see Darnell, 1982, and section 3.10 below). Moreover, cordycepin blocks mRNA appearance in the cytoplasm, perhaps due to its inhibition of poly(A) synthesis. Alternatively, this effect may result from very rapid degradation of the poly(A)$^-$ messengers on reaching the cytoplasm. However, some poly(A)$^-$ mRNAs (e.g. histone messengers) are clearly processed, selected and exported efficiently in the absence of a poly(A) tag. Some mitochondrial transcripts carry short poly(A) tracts, although they remain confined within the mitochondrion. Finally, many of the stored cytoplasmic messengers found in sea urchin oocytes undergo extensive polyadenylation following fertilisation (Slater & Slater, 1974), although elsewhere poly(A) addition is confined to the nucleus. Overall, poly(A) may serve several functions in different situations, but none of these is universally dependent on its presence.

The size of cytoplasmic mRNAs ranges from 8S up to about 20S, though larger individual messengers are known, such as those coding for myosin heavy chains (26S), thyroglobulin (33S) and vitellogenin (>30S). On average, mRNAs are some four to five times smaller than denatured hnRNA chains (2000 as against 8–10000 nucleotides). The stability of messenger RNAs varies considerably. Some are extraordinarily stable, such as globin mRNA in mammalian reticulocytes or crystallin mRNAs in vertebrate lens fibres – where translation continues long after nuclear breakdown and cessation of transcription. In cultured liver cells, Stiles *et al.* (1976) have measured different half-lives for alanine aminotransferase mRNA (12–14 hours) and for tyrosine aminotransferase mRNA (2 hours). Thus long and short-lived messengers can coexist within the same cell (see also Berger & Cooper, 1975). The same messenger species may also show different stabilities during successive stages of differentiation; for instance, crystallin mRNAs are shorter-lived in lens epithelial cells than in lens fibre cells.

Messenger RNAs do not show the sequence interspersion pattern characteristic of both DNA and hnRNA. Sequences representing repetitive DNA are found in most cytoplasmic mRNA populations (Spradling *et al.*, 1974), but these are due entirely to a sub-population of messengers derived from repetitive genes. The other mRNAs present are all of single-copy origin (transcribed from unique genes), with no detectable interspersion of repetitive sequences (see Klein *et al.*, 1974). One exception to this rule occurs among the cytoplasmic RNAs of mature oocytes and early embryos, where long transcripts containing interspersed repetitive sequences have been found both in *Xenopus* and in sea urchins (§4.9 and Davidson *et al.*, 1982). Most mRNAs are considerably longer than would be required merely to code for their known protein products. Nevertheless, mRNAs are generally monocistronic, i.e. only one polypeptide chain is encoded by each messenger chain. By contrast, some nuclear mRNA precursors are polycistronic, but are processed to monocistronic messengers (e.g. adenovirus late RNAs in virus-infected cells). The excess sequences in mRNA are present as 5' and 3' *untranslated* regions, which probably function in ribosome binding (5' end) and in translational termination (3' end). However, the large size of these regions in many messenger RNA species suggest that they may fulfil other functions, as yet uncharacterised.

3.8 Split genes

In procaryotic systems, each gene is *colinear* with its product. That is to say, the entire gene is transcribed into mRNA, and each triplet of bases between the initiating (AUG) and terminating (nonsense) codons of the messenger is then translated into the corresponding amino acid, according to the genetic code. Thus a typical bacterial gene is one continuous stretch of DNA specifying a single polypeptide product. *A priori*, there was no reason to doubt that this would also prove true for eucaryotes.

Early models relating large hnRNA precursors to smaller mRNA products assumed that a long stretch of each hnRNA chain would be discarded during processing, leaving an mRNA fragment probably derived from one end. Since 3' poly(A) was discovered some years before 5' capping, one widely accepted model involved conserving a 3' messenger sequence plus attached poly(A), with 5' sequences from the original precursor being removed and degraded within the nucleus (see e.g. Georgiev, 1969). The discovery of 5' capping implied that 5' termini might also be conserved; however, either 5' caps (see §3.7.2 above) or 3' poly(A) (Derman & Darnell, 1974) could be added post-transcriptionally onto cleaved fragments of hnRNA.

The first indication that things might be otherwise in eucaryotes came from viral systems, where certain messenger RNAs are not colinear with their genes as mapped by sequencing and restriction analysis of the viral DNA. With the advent of cloned genomic DNAs in the later seventies, it became apparent that many eucaryotic messengers are also non-colinear with their genes. Several lines of evidence pointed towards this conclusion. Anomalous restriction patterns were often observed within gene regions; these involved both extra cleavage sites not predicted from RNA or protein sequence data, and also much greater-than-expected separations between some predicted sites.

A more striking demonstration came from R-loop analysis, i.e. from EM examination of the hybrids formed between an mRNA and its corresponding cloned gene. Under appropriate hybridisation conditions, the mRNA strand displaces the non-coding DNA strand along the entire length of complementary sequence, forming an RNA–DNA hybrid plus an *R-loop* of single-stranded DNA spanning each coding region. Co-linearity would predict a single R-loop per gene. In practice, however, one commonly finds two or more R-loops, separated by duplex DNA loops where the mRNA has *not* displaced one strand. Presumably this is because the intragenic sequences (introns) present in the duplex DNA loops are not represented in the mRNA chain. This is readily seen under EM, where duplex structures (DNA–DNA or RNA–DNA) are much thicker than single-stranded (R-loop) structures. An example is shown in fig. 33, based on the mouse β globin gene analysed by Tilghman *et al.* (1978a). One major and one minor duplex DNA loop (termed *introns*) split this gene into three coding segments (exons) represented by R-loops.

The power of the R-loop method is illustrated in fig. 34, showing the complex pattern of loops generated by hybridising the 34-piece vitellogenin gene to its messenger RNA (Wahli *et al.*, 1980). In this case, the displaced R-loop extends across the entire transcription unit, giving a pattern of thicker RNA–DNA hybrid regions punctuated by thinner loops of single-stranded intron DNA (most of these have not reformed duplexes with the displaced DNA strand under the conditions used, which makes loop counting easier!).

The implication, both of these multiple R-loops and of the anomalous restriction maps, is that many eucaryotic protein-coding genes are *split*.

A Electron microscope view of R-loop structure

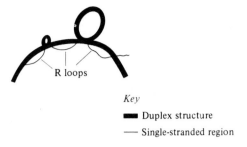

B Interpretation of R-loops

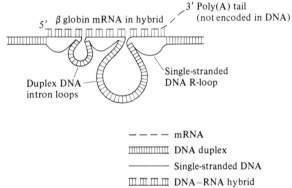

C Implied structures of gene and mRNA

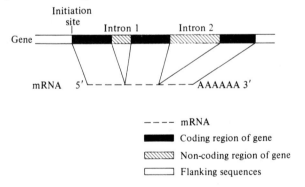

Fig. 33 R-loop analysis of β globin gene (based on Tilghman *et al.*, 1978a).

That is to say, the coding sequence is not continuous, but is interrupted by one or more *intron* (intervening) sequences; these introns are *not* represented in the final mRNA and are therefore non-coding. Thus each such gene is a mosaic structure, alternating between mRNA-coding *exon* sequences and non-coding *intron* sequences along its length. Before

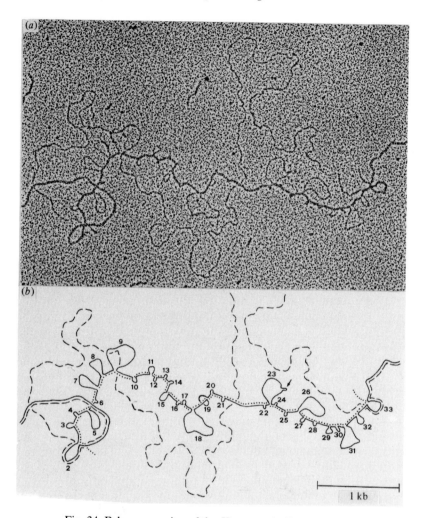

Fig. 34 R-loop mapping of the *Xenopus* vitellogenin A1 gene, by hybridisation with the 6.3kb mRNA derived from it. Part (*a*) electron micrograph; part (*B*) diagram of loop pattern; bar represents 1kb. For the most part, the displaced DNA strand has not reformed duplex structures with the non-hybridised intron loops, which are numbered from 1 to 33 in part (*b*). ———, coding DNA strand; _ _ _ _ _ _, non-coding (displaced) DNA strand; · · · · ·, mRNA strand. Photographs kindly supplied by Prof. W. Wahli (University of Lausanne, Switzerland) and reprinted with permission from the copyright holders, *Cell*. From W. Wahli *et al*. (1980) *Cell* **20**, 107–17.

considering in detail how a continuous mRNA chain can be derived from a split gene, the occurrence of introns in eucaryotic genes will be reviewed briefly.

Introns are characteristic of many but not all eucaryotic protein-coding genes. Most histone genes are intron-free (i.e. colinear with their mRNA products); however, there are some intron-containing minor histone genes in chickens. Introns are absent from many other mRNA-producing genes in lower eucaryotes and in *Drosophila*, though some have been found even in these systems, for instance in the actin genes of *Drosophila* and yeast. But introns occur much more commonly in the genes of animals with larger genome sizes (table 1). Some but not all tRNA-coding genes are also split by small introns located adjacent to the anticodon loop sequence (see §3.6 above), while the 5S rRNA genes are intron-free. In most higher eucaryotes the repetitive genes coding for 18S, 5.8S and 28S rRNAs also lack introns. In *Drosophila*, however, a considerable proportion of the 28S genes contain one intron each; several distinct classes of intron sequence have been described at this site in different rDNA repeat units (Dawid and Wellauer, 1977). It seems probable that these split 28S genes are not expressed, thus the functional rDNA units in *Drosophila* conform to the more usual colinear pattern for rDNA. In lower eucaryotes, by contrast, introns are found in *all* copies of the rDNA unit, again located in the 28S gene (one such intron in *Tetrahymena*, two in *Physarum*). Thus the presence of introns in rDNA does not in itself prevent expression. The removal of intron sequences from rRNA precursors in these systems involves a 'self-splicing' mechanism dependent on features of the intron itself; this pathway is distinct from that involved in tRNA or mRNA splicing (see e.g. Cech, 1983).

One case where introns occur in a procaryotic system has recently been described; short introns are located in the anticodon loop regions of two tRNA genes in *Solfolobus solfactaricus*, a member of the Archaebacteria (Kaine *et al.*, 1983). However, no introns have been detected in any gene among members of the other main procaryotic group, the Eubacteria (which includes *E. coli* and most other widely studied bacteria). Thus tRNA splicing, at least, appears very ancient in evolutionary terms.

Intron sequences in eucaryotic mRNA-coding genes range in size from about 60 bp up to several thousand bp, in contrast to the much shorter (10–30 bp) introns of tRNA genes. The number of introns within different genes also varies widely; e.g. two in each of the globin genes, seven in the chick ovalbumin gene, 16 in the chick conalbumin gene, 33 in the *Xenopus* vitellogenin gene, and over 50 in the chick collagen Iα2 gene. Introns are mainly unique sequences, but in some cases include repetitive elements (Ryffel *et al.*, 1981); the presence of the latter within single-copy protein-coding genes may in part account for the interspersed sequence arrangement of many hnRNAs (containing intron transcripts, see §3.9) but not of the mRNAs derived from them.

3.9 RNA splicing

How is a continuous messenger chain derived from a split gene composed of coding exons and non-coding introns? Two possibilities may be envisaged: either (i) RNA polymerase II transcribes only the exons, somehow passing over the introns; or else (ii) the initial pol II product contains transcripts of both exon and intron sequences, the latter being excised during RNA processing and the former joined together (spliced) to create a continuous mRNA chain.

A clear answer to this question was obtained in the β globin system (see figs. 33 and 35). A 15S hnRNA precursor containing globin mRNA sequences had previously been identified in erythroid nuclear RNA from both avian (MacNaughton *et al.*, 1974) and mammalian (Ross, 1976) sources under denaturing conditions. Curtis *et al.* (1977) identified this 15S species as the largest detectable precursor to β globin mRNA, while α globin mRNA was derived from a smaller nuclear precursor of about 11S. The length of the 15S precursor (1500 to 1900 nucleotides in different mammalian species) is about three times that of the 9S β globin messenger, and is close to that estimated for the genomic β globin gene by summing the lengths of the R-loops and intron loops observed in mRNA–DNA hybrids under EM (see fig. 33). When the R-loop analysis was repeated with purified 15S precursor RNA in place of β globin messenger, only a single R-loop was obtained, with no detectable intron loops (Tilghman *et al.*, 1978b). This implies that the 15S precursor is colinear with the entire β globin gene (fig. 35).

Analysis of nuclear RNA precursors leads to a similar conclusion for other split genes. In all cases, the large primary transcript represents both intron and exon sequences, of which only the latter are conserved to form mRNA, the former being degraded within the nucleus (e.g. for ovalbumin gene transcription, see Tsai *et al.*, 1980). Thus transcription itself is colinear, but RNA processing removes the intron sequences from *internal* regions of the transcript to generate mature mRNA.

This raises two important questions concerning the RNA processing system. How are intron transcripts distinguished from exon transcripts within a long hnRNA chain? And how are the exon transcripts joined together in the correct order to form a continuous mRNA chain? One

A   **Electron microscope view of R-loop structure**

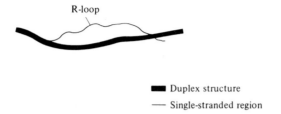

R-loop

■ Duplex structure

— Single-stranded region

B   **Interpretation of R-loop**

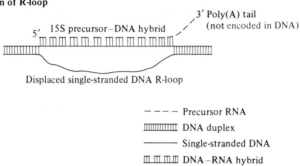

3′ Poly(A) tail
(not encoded in DNA)

5′   15S precursor–DNA hybrid

Displaced single-stranded DNA R-loop

– – – – Precursor RNA

⫿⫿⫿⫿⫿ DNA duplex

———— Single-stranded DNA

⫿ ⫿ ⫿⫿ DNA–RNA hybrid

C   **Implied relationship of mosaic β globin gene to mRNA and 15S precursor.**

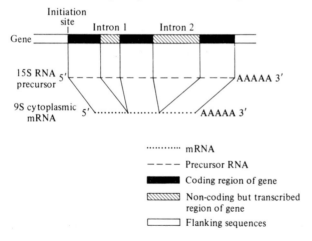

Initiation
site   Intron 1   Intron 2

Gene

15S RNA 5′
precursor   AAAAA 3′

9S cytoplasmic 5′
mRNA   AAAAA 3′

············ mRNA

– – – – Precursor RNA

■ Coding region of gene

▨ Non-coding but transcribed
region of gene

▭ Flanking sequences

Fig. 35 R-loop analysis of 15S β globin precursor RNA (based on Tilghman *et al.*, 1978b).

simple model, which could kill both birds with one stone, proposed that introns might be flanked by inverted repeat sequences in the DNA, such that hnRNA transcripts would form foldback duplex stems and single-stranded intron loops. The latter could be removed by an appropriate loop-specific endonuclease, leaving the exon ends ready for splicing by an RNA ligase (fig. 36).

Definitive proof for such a model would require that all introns be flanked by inverted repeat sequences in the gene. Sequence analysis of twelve boundary regions flanking six of the seven introns in the chicken ovalbumin gene failed to identify such inverted repeats. However, common sequence features were found at the six exon–intron junctions, and others at the six intron–exon junctions (Catterall *et al.*, 1978). These same *consensus sequences* are found (with minor variations) at the correspond-

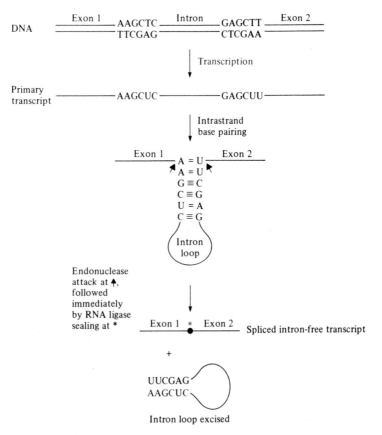

Fig. 36 Inverted repeat model for removal of intron transcripts (now discounted).

ing boundary sites in all other mRNA-coding genes containing introns. The consensus sequences in the DNA are as follows: $5'\text{-}^A_CAG \downarrow GT^A_GACT\text{-}$intron$(Py)_6XCAG \downarrow G^G_T\text{-}3'$, where $\downarrow$ marks the splice points (Cech, 1983); the GT and AG dinucleotides at the 5' and 3' ends of the intron appear to be invariant in all mRNA-coding genes. These features are not, however, found in any tRNA-coding genes, whose introns (when present) are much shorter, nor in the rRNA-coding genes of lower eucaryotes (see Cech, 1983).

The discovery in 1980 (Lerner *et al.*, 1980; Rogers & Wall, 1980) that the 5' sequence of U1 snRNA is partially complementary to *both* sets of consensus sequences flanking the introns in mRNA-coding genes, provided the basis for an alternative model of RNA splicing. As mentioned previously (§1.9), the snRNAs are stable, abundant and confined to the nucleus, where many occur base-paired to large hnRNA molecules. They are also present in ribonucleoprotein complexes (snRNPs) which can be precipitated from nuclear extracts by antibodies from patients with the autoimmune disease *lupus erythaematosus*. These features are consistent with – but do not prove – a role for snRNAs in mRNA splicing from hnRNA precursors. Recently, highly specific antibodies against U1-containing snRNP particles have been shown to inhibit the splicing process in a soluble *in vitro* system, whereas antibodies against RNP particles containing other types of snRNA have no such effect (see Padgett *et al.*, 1983). This suggests that U1 snRNA is indeed essential for splicing of mRNA. The sequences involved, together with an outline of the proposed splicing mechanism, are given in fig. 37 (after Rogers & Wall, 1980).

Since U1 snRNA can only form base pairs with the intron consensus sequences, it cannot hold the exon ends together after removal of the intron loop. Thus splicing by an RNA ligase must follow on immediately after endonuclease cutting, otherwise free exon ends might be rejoined in an incorrect order (a more serious problem for complex multi-intron hnRNAs). This might be achieved within a single macromolecular complex containing both ligase and endonuclease activities together with U1 snRNA as a recognition component. In fact, U1-containing snRNPs seem to contain as many as eight different protein species. Recent evidence suggests that introns may be excised in an unusual lariat form (with the 5' end of the intron linked to an adenosine residue near its 3' end), rather than as linear or circular RNA fragments (see Weissmann, 1984). Apparently the 5' end of the intron is cleaved first, as a transient intermediate contains the intron lariat still attached to the 3' exon.

Some variability is found among the consensus sequences identified at splice points in different mRNA-coding genes, though all can base pair to

some extent with U1 snRNA. Cases in which the sequence matching is relatively poor could result in optional splicing, i.e. sites which are not always recognised by the processing system. A case of this type has been identified in the mouse lens system, where a variant form of αA crystallin

A   **Consensus sequences at intron/exon boundaries**
     (given as for the hnRNA transcript, but based on sequence analysis of the gene)

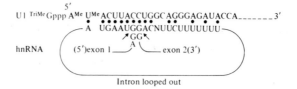

B   **5' end of U1 snRNA sequence**

$$^{5'}_{\text{TriMe}}\text{Gppp A}^{\text{Me}}\text{ U}^{\text{Me}}\text{ ACUUACCUGGCAGGGAGAUACCA} \text{_____ 3'}$$

C   **Proposed matching for splicing**

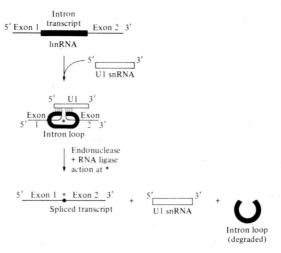

Fig. 37 Splicing mechanism proposed by Rogers & Wall (1980).

containing an insert of 23 extra amino-acids (αA^{Ins}) results from optional splice points within a long (1376 bp) intron in the gene. Sometimes these splice sites are recognised in the hnRNA as delineating an extra 69 bp exon which codes for the insert; more often, however, this sequence is not recognised as an exon, leading to removal of the insert-coding sequence along with the rest of the long intron (fig. 38). Thus two messenger RNAs can be derived from a single gene by alternative splicing pathways (see King & Piatigorsky, 1983).

In several other systems, alternative splicing pathways are subject to developmental control. During B lymphocyte differentiation, immuno-globulin M expression shifts from a membrane-bound to a secreted form, a difference reflected in the C terminal sequences of the component heavy chains (μ_m and μ_s respectively). μ_m and μ_s mRNAs are identical apart from their 3' termini, which encode respectively a 41 amino-acid hydro-phobic region (characteristic of trans-membrane proteins) and a 20 amino-acid hydrophilic region (characteristic of secreted proteins). Both μ_m and μ_s mRNAs are derived from the same μ gene by alternative RNA processing pathways (Rogers *et al.*, 1980; Early *et al.*, 1980b). The immunoglobulins are further discussed in section 4.2.

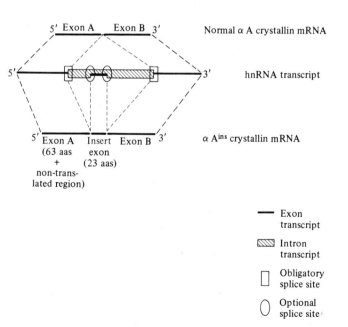

Fig. 38 Alternative splicing pathways for mouse αA crystallin hnRNA (based on King & Piatigorsky, 1983).

Splicing sites occur commonly within intron sequences as well as at intron/exon boundaries, and in many cases the removal of a given intron transcript from hnRNA is accomplished in two or more stages (see e.g. Avvedimento *et al.*, 1980, on the processing of chick collagen Iα2 RNA). Furthermore, different intron sequences are removed from a given type of hnRNA precursor in an obligatory or at least a preferred order (see e.g. Tsai *et al.*, 1980, on the processing of chick ovomucoid and ovalbumin RNAs).

A typical case in point is the rabbit β globin transcription system, analysed in detail by Grosveld *et al.* (1981b). The largest spliced intermediate among β globin hnRNAs has lost only 40 nucleotides from the smaller intron sequence, while the next largest species has lost the entire 126 bases of this intron (in both the large intron sequence remains intact). A further intermediate retains only 90 nucleotides of the large (573 base) intron sequence. Thus the two intron sequences are processed sequentially (small before large), and both are removed from the hnRNA precursor in two stages (fig. 39).

As to the order of RNA processing events within the nucleus, 3′ poly(A) is added soon after transcription (Salditt-Georgieff *et al.*, 1980), and apparently precedes splicing. However, poly(A) is not a necessary signal for splicing, since this process can continue even when polyadenylation is blocked by cordycepin (Zeevi *et al.*, 1981). The overall pattern of hnRNA conversion into mRNA is dealt with in section 3.10 below.

Before leaving the subject of introns and RNA splicing, some mention must be made of their possible evolutionary significance. The intronless pattern characteristic of Eubacterial systems is obviously more efficient in energy terms, both for DNA replication and for transcription (because fewer energy-rich DNA and RNA precursors are 'wasted' on non-coding

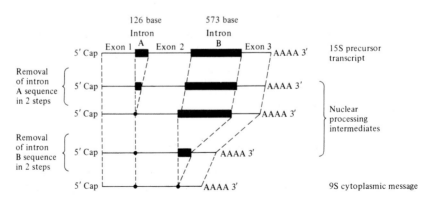

Fig. 39 Splicing of 15S β globin precursor RNA (after Grosveld *et al.*, 1981b).

sequences). Presumably eucaryotes derive other benefits from their huge genomes, split genes and apparently wasteful transcription habits!

The introns which occur at corresponding positions in related genes are found to diverge widely both in length and in sequence, much more so than the corresponding exons (cf. the contrast between spacer and coding sequences in tandemly repeated genes; §1.8). This is shown clearly in a family of four related vitellogenin genes in *Xenopus* (Wahli *et al.*, 1980; §6.4), and in numerous studies of e.g. globin gene evolution between species (van den Berg *et al.*, 1978). Thus introns may provide reservoirs of rapidly evolving sequences, which are not themselves performing a coding function and hence are not under direct selective pressure. This variability could be used (over evolutionary time) to supplement the more conservative exon repertoire, via alternative splicing pathways etc.

A more radical suggestion was proposed by Gilbert (1978) and amplified by Blake (1978); in essence, they asked whether 'genes-in-pieces' might imply 'protein-in-pieces'. Many proteins are composed of several functional 'domains', i.e. separately-folded regions of the polypeptide chain responsible for different aspects of that protein's function. If these domains are encoded by different exons of the gene, then a novel and rapid mechanism for protein evolution can be envisaged. This would involve shuffling exon units into new arrangements by recombination events within intron sequences. New split genes generated in this way could give rise to multifunctional (and potentially useful) proteins, since the exon-coded domains would already have independent functions. Thus a preexisting protein-coding gene could acquire new functions (or lose old ones) far more rapidly than by simple point mutation. Though speculative, this idea has received some experimental support, to the extent that the exons in many genes do indeed code for distinct functional domains of the protein chain (see e.g. Sakano *et al.*, 1979; Craik *et al.*, 1980; Jung *et al.*, 1980; Stein *et al.*, 1980). However, this is not invariably the case; thus the exons in a nematode myosin gene do not delineate the known protein domains (Karn *et al.*, 1983). Recently, Blake (1983) has argued that a one-to-one correlation between gene exons and protein domains is unnecessary; possibly a small group of exons might encode a single functional domain.

3.10 mRNA and hnRNA populations

The primary transcripts from split genes range from three to six fold longer than their final mRNAs. If all hnRNA chains were mRNA precursors, then around 20% of nuclear RNA sequences should be exported to the cytoplasm. In fact, the observed proportion is much lower

Table 3. *Sequence complexity of hnRNA and mRNA populations (data from Lewin, 1980)*

	Cell type			
	Sea urchin gastrula	Sea urchin adult intestine	Mouse brain	Mouse Friend cells (erythroid cell line)
hnRNA complexity[a]	16–17%	18%	15–20%	5–6%
mRNA complexity[a]	1.4%	0.5%	3.5%	1.2%

[a]Complexity is expressed as the percentage of total single-copy DNA sequences represented (at any level) in the RNA population.

than this ($\sim$2% in erythroid cells; Tobin *et al.*, 1978). This discrepancy could be explained in two ways (not mutually exclusive). Either (a) all types of hnRNA are processed inefficiently to messengers, in which case the low rate of conversion is simply a quantitative effect. Alternatively (b) only certain types of hnRNA are selected for processing to messengers, in which case a qualitative control is being exerted, and those hnRNAs which turn over within the nucleus will be mostly different from those destined to form mRNAs. In either case, the complexity of an hnRNA population should considerably exceed that of the corresponding mRNA population, though more so if explanation (b) were correct. Estimates of RNA sequence complexity can be derived by RNA excess hybridisations (§3.2.4) with isolated single-copy DNA (see §1.5) as the labelled tracer. As shown in table 3, nuclear hnRNA populations are at least five-fold more complex than cytoplasmic messenger populations from the same cell types.

The assumptions underlying these complexity estimates are somewhat involved, and the figures quoted are based on Lewin's recalculation of original data from several authors (see chapters 24 and 25 in Lewin, 1980). It should be noted that these estimates are based on plateau saturation values; this means that *all* RNA sequences present – however rare – will contribute to the final complexity measurement, because the RNA component is in vast excess during hybridisation. Thus the saturation approach cannot distinguish between abundantly and rarely transcribed RNA sequences.

To make this distinction an alternative kinetic approach has been devised. This involves reverse transcribing a poly(A)$^+$ RNA population into complementary cDNA strands, and then hybridising these labelled

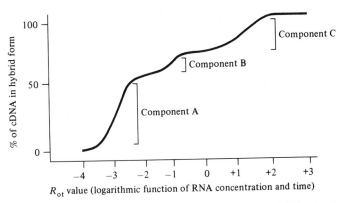

Fig. 40 Oviduct cDNA hybridisation to its template mRNA population (based on Axel *et al.*, 1976).

cDNAs with an excess of the template RNA population. As mentioned above (§3.2.4), this experiment is analogous to DNA renaturation, in that it provides an estimate of sequence abundance and complexity within each kinetic class. Fig. 40 shows such an analysis using cDNA prepared from hen oviduct poly(A)$^+$ mRNA.

Component A in fig. 40 is of low complexity (about 2 kb), and corresponds to a single highly abundant messenger representing some 50% of the total mRNA mass (about 100 000 copies per cell). Presumably this messenger encodes ovalbumin, which is by far the most abundant protein product in hen oviduct cells (see chapter 6). Component B is of higher complexity (14 kb), corresponding to about 7 mRNA species of similar size to the ovalbumin message; between them these represent some 15% of the mRNA mass, and each is present in about 4000 copies per cell. This component probably includes mRNAs coding for the other major egg white proteins, i.e. ovomucoid, conalbumin, lysozyme etc. Component C is much more complex (25 000 kb), comprising about 12 500 different mRNA sequences of the same average size; each is present in only 5 copies per cell on average, but between them they total around 35% of the mRNA mass.

The oviduct pattern described above (Axel *et al.*, 1976) is typical of mRNA populations derived from highly specialised animal cells. One or a few messengers encoding the principal cell product(s) are extremely abundant. A larger range of moderately prevalent mRNAs, each present in fewer copies per cell, code for the other major cellular proteins. Finally there is a much more heterogeneous group of rare mRNAs each occurring in only a few copies per cell. The functional significance of this last group is not entirely clear. Those present in perhaps a few tens of copies per cell probably code for metabolic enzymes and other proteins required in small amounts for cell survival. The term 'housekeeping proteins' aptly describes this category, and in some tissues their mRNAs form a discrete kinetic component.

Other messengers, however, are much rarer, sometimes occurring in one or fewer copies per cell. These mRNAs might of course be present at significant levels in a minority of cells but completely absent in others (i.e. cellular heterogeneity). Galau *et al.* (1977) have argued that even very rare mRNAs may nevertheless code for vital cellular proteins (including RNA polymerase subunits), based on estimates of turnover and synthesis rates for these proteins in liver cells. However, not all rare-class cytoplasmic mRNAs are translated into proteins. In duck erythroblasts, only 200 rare messengers occur on the polysomes, while some 1200 rare messengers are non-polysomal and apparently excluded from translation (Imaizumi-Scherrer *et al.*, 1982).

Yet another possible explanation is that very rare mRNAs result from leaky transcriptional and/or processing/export controls. Very low levels of globin mRNA sequences have been reported in nuclear and cytoplasmic RNA populations from several non-erythroid tissues, including brain, kidney and cultured fibroblasts (Humphries *et al.*, 1976). But in several other systems, the genes coding for specialised cell products are transcribed *only* in the appropriate cell type (e.g. for liver, see Derman *et al.*, 1981), their transcripts being undetectable in other tissues. If generally true, this would argue against 'leaky' control systems.

It must be emphasised that current protein-separation techniques are totally inadequate to resolve all the components of a cellular protein population representing the full mRNA diversity of that tissue (commonly 5–30×10^3 different messengers). Thus even the best two-dimensional (2D) protein gels probably show only the most abundant 10% of proteins in any particular tissue (see Paul, 1982). It is by no means improbable that several thousand different mRNAs might be required to code for the vast range of enzymes, membrane proteins, cytoskeletal elements, nuclear proteins etc. expressed even in highly specialised cell types *in vivo*. Many of these proteins will be needed in most if not all cell types, resulting in significant overlaps between the mRNA populations of different tissues (as shown by competition hybridisation experiments). These overlaps are largely confined to the rare or moderately prevalent mRNA classes coding for housekeeping proteins; the abundant messengers encoding 'luxury proteins' are more likely to be cell-type specific. However, messengers which are abundant in one cell-type can sometimes be detected among the moderately prevalent or rare-class mRNAs in another. One case in point is the occurrence of crystallin mRNAs (abundant in the lens) at low levels in the RNA of certain other tissues in chick embryos (Clayton *et al.*, 1979; Agata *et al.*, 1983).

In cases where hnRNA populations have been compared by competition hybridisation, the overlaps between tissue types generally predominate over any differences. Thus the hnRNA populations from sea urchin blastula and pluteus embryos comprise virtually the same set of sequences (Kleene & Humphries, 1977). In rat, some 16%, 11% and 5% of the single-copy DNA sequences are expressed in the hnRNA populations of brain, liver and kidney respectively (Chikaraishi *et al.*, 1978; recalculated by Lewin, 1980). But these populations appear to be 'nested'; in other words, most of the kidney hnRNAs are a subset of the liver hnRNA population, which in turn is largely a subset of the brain hnRNAs (Chikaraishi *et al.*, 1978). This rather gross analysis (through competition hybridisation) does not preclude tissue-specific expression of certain hnRNAs and messengers, for instance in liver but not in brain (see Derman *et al.*, 1981). However, it does suggest that the tissue-specific component is a relatively small proportion of the total hnRNA population. In general, hnRNA populations are more similar in sequence composition than are the mRNA populations derived from them. Presumably this means that many hnRNA species turned over within the nucleus are qualitatively the same in all cell types.

These arguments suggest a negative answer to the original question posed in this section, as to whether a cellular mRNA population is processed from *all* types of hnRNA transcribed in the nucleus. One could, indeed, envisage an extreme system of eucaryotic gene control in which *all* available genes are transcribed into hnRNAs, the bulk of which would be turned over within the nucleus, while only a selected minority would be processed to mRNA for export into the cytoplasm (see §4.9 below). Wold *et al.* (1978) found that most of the single-copy DNA sequences expressed in sea urchin blastula mRNA are also represented in the hnRNA population of adult intestine nuclei, but not in the cytoplasmic mRNA of these intestine cells. This example suggests that gene expression is subject to important qualitative controls at the level of RNA processing and/or export. Similar controls may also be exerted in quantitative terms. In a study using cloned probes to assay the levels of 9 specific RNA sequences expressed in cultured Chinese hamster ovary (CHO) cells, the relative concentrations of these RNAs were found to vary by up to 10 fold in nuclear RNA, but by as much as 100 fold in the cytoplasmic mRNA population (Harpold *et al.*, 1979). However, this could simply reflect different mRNA stabilities in the cytoplasm, rather than differential processing and/or export.

On the other hand, it is clear that the proportion of single-copy sequences represented in hnRNA varies considerably between different cell types (see e.g. Chikaraishi *et al.*, 1978; discussed above). Thus all genes cannot be expressed in all tissues. There is also strong evidence for transcriptional controls governing the expression of tissue-specific genes. Several liver-specific mRNA sequences are detectable only in liver hnRNA but not e.g. in brain hnRNA (Derman *et al.*, 1981), as assayed by hybridisation with cloned probes. Similarly, Ernst *et al.* (1979) have detected a small proportion of sea urchin single-copy sequences which are expressed in adult intestine hnRNA but not in gastrula hnRNA (see §4.3). This could represent the switching-on of several hundred tissue- or stage-specific genes, again implying a measure of transcriptional control. Many repetitive sequence transcripts which are confined to the nucleus (i.e. present in hnRNA but not mRNA) are also expressed in a tissue- and stage-specific pattern (Scheller *et al.*, 1978).

As discussed above, only some types of hnRNA are processed to messengers for export, many other species being retained and turned over within the nucleus. In cultured CHO cells, only 25% of the capped hnRNA chains are found to carry poly(A), but this fraction contains some 70% of specific mRNA sequences (identified using several cloned probes). Thus a majority of capped hnRNA chains probably contain neither 3' poly(A) nor sequences destined to form messenger (Salditt-Georgieff & Darnell, 1982); this implies that polyadenylation may distinguish messenger precursors from other hnRNA species (i.e. those destined for turnover within the nucleus).

Gene control is thus exerted at a variety of levels in eucaryotic systems (Darnell, 1982); these include the transcription, processing, export, degradation, storage and selective translation of RNAs. As for the poly(A)$^-$ hnRNA chains turned over in the nucleus and not serving a messenger-precursor function, their significance remains speculative, although a major regulatory role has been proposed (see §4.9).

4

Molecular strategies in development

Summary

This chapter broaches the complex question of how genes act together in a coordinated fashion during animal development. Three fundamental mechanisms might underlie the process of differentiation, whereby particular groups of cells acquire specific functions not expressed significantly in other cell types. These are: (i) amplification of genes whose products are required at high levels; (ii) elimination of genes whose products are not required at all; or (iii) selective expression of different sets of genes in different tissues, the nuclear genome remaining constant throughout. Although specific cases of amplification and elimination are known (§4.2), and DNA rearrangements define the antibody gene expressed in particular clones of B lymphocytes (§4.2), these appear to be exceptions rather than the rule in eucaryotic systems. Section 4.2 also discusses some of the evidence on nuclear totipotency in animals, i.e. whether nuclei from differentiated cells retain the ability to direct a complete programme of development.

Section 4.3 deals with changes in overall RNA populations during embryonic development and tissue differentiation in sea urchins. A complex array of RNA sequences and other reserve materials is inherited by the zygote from the oocyte; the accumulation of these reserves during oogenesis is summarised in section 4.4. They are utilised by the embryo after fertilisation, i.e. during early cleavage development (zygote $\rightarrow$ blastula), which process is consequently under maternal control at least in part (§4.5). Gradually, however, the embryo's own genome assumes control over development, usually by the gastrula stage, though earlier in slowly-cleaving mammalian embryos (§4.5).

111

In mosaic development (e.g. in molluscs or ascideans), maternal information in the form of 'morphogenetic determinants' is distributed differentially between different regions of the zygote cytoplasm, and becomes partitioned into distinct groups of cells during cleavage, so directing those cells along different paths of differentiation (§4.6). Though many examples imply the existence of such morphogenetic determinants, their chemical nature and mode of action remain largely unknown. Despite superficial contrasts, the regulative mode of development shown by other animal groups is not fundamentally very different (except perhaps in mammals). In most regulative embryos, the morphogenetic determinants do not become segregated into different groups of cells until later in cleavage, and there seems to be some compensating mechanism which can reestablish a normal balance of determinants even when parts of the embryo are deleted (§4.6).

Some molecular correlates of the earliest signs of differentiation (i.e. emergence of different cell types) are summarised in section 4.7. During later development, cells often become committed (determined) to follow a particular course of differentiation well before they begin to express the differentiated products (markers) characteristic of that pathway. The roles of short-range inductive interactions and long-range hormonal influences, either initiating or modifying the course of tissue differentiation, are considered briefly in section 4.8. Finally, a model for eucaryotic gene control proposed by Britten & Davidson in 1969, is discussed in some detail (§4.9). In one or more of its several manifestations, this model can account for many peculiar features of eucaryotic genome organisation, transcription and development.

4.1 Introduction

The preceding chapters have covered general features of animal genes and their expression. Before exploring a few systems in greater detail, it is appropriate to consider how genes might work in concert during development.

Our picture of animal development is strongly coloured by the organisms most widely studied – namely sea urchins, *Drosophila*, *Xenopus*, chicken, rodents, and more recently the soil nematode *Caenorhabditis elegans*. Only in *Drosophila* is our developmental knowledge complemented by detailed genetics, though the nematode system could soon become a rival in this respect (Brenner, 1974). Sea urchins are virtual non-starters in genetic terms, while among vertebrates the mouse has a clear advantage, with many inbred laboratory strains. Of course, the absence of conventional genetics does not preclude molecular studies through gene cloning, RNA population analysis etc., as for instance in sea urchins. However, such studies cannot easily focus on gene functions with defined effects during development. This option is only widely available in *Drosophila*, whose developmental patterns are in some respects unusual. For this reason, insect systems are considered separately in chapter 7.

This chapter will deal with molecular aspects of several questions raised by classical studies of animal development. Is the genome identical in all cells of an organism, or are there irreversible changes in the nuclear DNA of particular cell types during their differentiation? To what extent are the RNA and protein populations of early embryos supplied from the fertilised egg? How is this developmental information built up during oogenesis, and for how long does it exert maternal control over embryonic development? When does the embryo's own genome become active, and when do signs of differential gene expression first distinguish one prospective cell type from another? Are regional differences built into the structure of the egg itself (as in mosaic development), or are they acquired during embryogenesis (as apparently occurs during regulative development)? What factors govern the selection and expression of tissue-specific genes during the determination and subsequent differentiation of particular cell types? And finally, can these features be accommodated by any plausible model for eucaryotic gene regulation?

The discussion below is mainly confined to early development, where the following sequence of events is applicable in all animals:

$$\text{Oogenesis} \xrightarrow{\text{fertilisation}} \text{zygote} \xrightarrow{\text{cleavage}} \text{blastula} \xrightarrow{\text{gastrulation}} \text{gastrula}$$
$$\xrightarrow{\text{organogenesis}} \text{tissue differentiation}$$

This whole field is covered much more fully by E. H. Davidson in his book 'Gene Activity in Early Development' (1976).

4.2 A constant genome?

A hallmark of metazoans is the specialisation of particular cell types to carry out different functions. Commonly this is reflected in the tissue-specific expression of certain abundant 'luxury' products, such as haemo-globin in vertebrate red blood cells, egg white proteins in the laying hen oviduct, or crystallins in the vertebrate eye lens. When the various cell types of a given animal are compared, their patterns of protein expression generally overlap in part but are otherwise distinctive (see Truman, 1974, 1982). This is true both in quantitative and in qualitative terms; each tissue is distinguished by its characteristic cellular levels, as well as by the presence or absence, of a wide range of housekeeping and luxury gene products. The process whereby cells acquire these distinctive patterns of gene expression is termed *differentiation* (§§4.7, 4.8).

As to the molecular strategy of differentiation, three basic mechanisms can be envisaged: (i) *amplification* of genes whose products are required in high abundance for a particular differentiation pathway; (ii) *deletion* of genes whose products are *not* required (at any level) for such a pathway; or (iii) *differential gene expression* from a constant genome.

Although (iii) appears to be the general rule in animal systems, examples of both amplification and deletion have been described in particular cases. A brief review of some salient evidence on genome constancy is given in the next three sections.

4.2.1 Amplification

Until recently, the only clear example in this category was the amplification of ribosomal genes (rDNA) during oogenesis (§1.8). Not all animals amplify these genes in their oocytes; in particular, those which undergo meroistic oogenesis (see §4.4) are able to accumulate huge reserves of ribosomes by a different mechanism.

rDNA amplification occurs during the pachytene stage of meiotic prophase, and probably involves copying one or more chromosomal sets of ribosomal genes via a 'rolling-circle' DNA intermediate (see fig. 41*A*; Hourcade *et al.*, 1973). In each *Xenopus* oocyte, between 1.5 and 2.5 million copies of the ribosomal repeat unit are produced in this way (Perkowska *et al.*, 1968), as compared to about 450 chromosomal copies per haploid genome. The amplified copies are in the form of free rDNA circles (nucleolar cores) each containing numerous repeat units. A single oocyte contains about 5000 such cores, and some 1500 *extrachromosomal nucleoli* are formed around them. These become engaged in active rRNA synthesis (see fig. 30) during later oocyte growth. The amplified rDNA copies are gradually broken down after use and lost during early embryogenesis; unlike their ribosome products they are not utilised in the embryo.

Another case of gene amplification during normal development has been reported in *Drosophila* egg chambers. The follicle cells secrete a proteinaceous chorion sheath around each egg prior to laying. Some 20 chorion proteins have been resolved (Waring & Mahowald, 1979), and their genes mapped to several different clusters in the genome. One such cluster contains four chorion protein genes spanning 6 kbp of DNA at site 66D on chromosome 3 (Griffin-Shea *et al.*, 1982). During the development of follicle cells, their chromosomes become *polytene*, i.e. several rounds of DNA synthesis occur without separation of the daughter duplexes, so that many DNA molecules come to lie side by side in perfect register (see also §7.1). But over and above this general increase in nuclear DNA content, the chorion protein genes become specifically amplified during egg laying (Spradling & Mahowald, 1980). The amplification process in this case probably involves limited DNA replication in the region of the chorion gene clusters (see fig. 41*B*), since adjacent DNA regions spanning some 40 kbp are found to be amplified to a lesser extent (decreasing outwards from the 66D site; Spradling, 1981). Multifork replication structures representing active chorion gene regions in *Drosophila* follicle cells have recently been visualised by EM spreading (Osheim & Miller, 1983). Interestingly, one of the genes at site 66D (but not the other three) is actively transcribed just prior to amplification, although the mRNA product is not at this stage translated into protein. Following amplification, all four genes in the cluster are actively transcribed, and their mRNAs are all translated into chorion proteins (Thireos *et al.*, 1980; Griffin-Shea *et al.*, 1982). Amplification of specific gene regions also occurs in the 'DNA puffs' of salivary gland chromosomes in the insect *Rhynchosciara* (Glover *et al.*, 1982).

A variety of cell types seem able to amplify particular DNA sequences when placed under selective pressure in culture. Chinese hamster ovary cells can metabolise low levels of the cytotoxic drug methotrexate (MTX) by means of the enzyme dihydrofolate reductase (DHFR), which is encoded by a single-copy gene. Sudden exposure to high levels of MTX is rapidly lethal to all cells in such cultures. If the MTX levels are increased gradually over an extended period in culture, however, a population of MTX-resistant cells is obtained. These cells contain levels of DHFR activity several hundred-fold higher than in normal (MTX-sensitive) cells, due to amplification of the DHFR gene by more than a hundred fold. In some cases, MTX-resistance is inherited stably over many cell generations, even in the absence of MTX itself. In other cases, MTX-resistance is rapidly lost from the progeny cells when MTX is withdrawn (unstable resistance). In the stably-resistant cells, an expanded homogeneously-staining region (HSR) of chromosome 2 has been identified, containing most of the amplified DHFR gene copies (Nunberg *et al.*, 1978). Moreover, a 135 kbp region of DNA becomes amplified in these cells; this is much larger than the DHFR gene itself, and may represent an entire *replicon* unit of DNA replication (Milbrandt *et al.*, 1981). The amplified sequences in an HSR are probably in a linear tandem array, which might arise from a multifork replication structure (cf. fig. 41*B*) by recombination between repetitive sequences (see Roberts *et al.*, 1983). By contrast, in unstably-resistant cells the amplified DHFR gene-copies are located on separate double-minute chromo-

A   Possible model for ribosomal gene amplification in young *Xenopus* oocytes

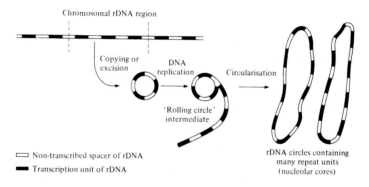

B   Chorion protein gene amplification in *Drosophila* egg chamber cells (after Spradling, 1981)

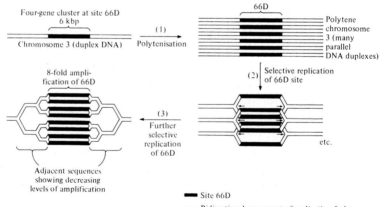

Fig. 41 Gene amplification.

somes, which are very small, lack centromeres and hence are easily lost during mitosis (Kaufman *et al.*, 1979).

One might expect such amplifications to be very rare events, detectable only because a single cell with the resistant (amplified) phenotype can survive the demise of many millions of sensitive (non-amplified) cells. However, recent data (Johnston *et al.*, 1983) suggests that amplification may occur quite commonly in cultured cells even in the absence of selective agents; spontaneous increases and decreases in DHFR gene number can be detected at relatively high frequencies (once per thousand divisions) in cultures never exposed to MTX.

Similarly, progressive increases in cadmium levels can be countered by gene amplification in a cultured mouse cell line. The Cd^{2+}-resistant cells contain abnormally high levels of metallothionein I (a protein which binds to and detoxifies heavy metal ions), due in part to amplification of the metallothionein I gene (Beach & Palmiter, 1981). Interestingly, metallothionein gene expression can be induced by glucocorticoid hormones as well as by heavy metal ions (e.g. Cd^{2+}). Recent deletion analysis of the 5′-flanking regions of a human metallothionein II_A gene has defined separate DNA elements involved in these controls. The sequences necessary for heavy-metal induction are duplicated at positions -38 to -50 and -138 to -150, while the sequence required for hormone induction is located between positions -237 and -268 (a region known to bind the hormone–receptor complex). These control elements appear to function independently (Karin *et al.*, 1984).

The ability to amplify DNA sequences *in vitro* does not necessarily imply that such mechanisms are used *in vivo* during normal differentiation. In several specific instances amplification clearly does *not* occur, even though the differentiated cell type may synthesise vast quantities of one or a few gene products. Thus globin genes are not amplified in erythroid cells (Bishop *et al.*, 1972), nor ovalbumin genes in chick oviduct cells, nor delta crystallin genes in chick lens (Zelenka & Piatigorsky, 1976). This was shown by mRNA or cDNA hybridisation (§3.2.4) to a vast excess of genomic DNA prepared from tissues in which the gene product is respectively abundant or absent. Both give the same hybridisation curve, implying a constant number of gene copies per haploid genome.

4.2.2 *Gene deletion and rearrangement*

The same evidence which rules out gene amplification in specific instances can equally be used to argue against a deletion model for differentiation.

If globin gene sequences are present at the same copy number in erythroid cell DNA as in, say oviduct DNA, then the process of oviduct differentiation clearly cannot involve deleting the globin genes, which will never be expressed in oviduct cells. Of course, such examples do not provide a formal proof of genome identity in all cell types; the number of possible permutations is simply too vast. But even if deletions occur occasionally, these arguments do at least preclude the possibility that *all* unexpressed genes might be lost routinely.

There are indeed cases where DNA is deleted during development. As long ago as 1899, Boveri noted that chromosome ends are lost from all somatic nuclei during early cleavage in *Ascaris*, complete chromosomes being retained only in the primordial germ cell nucleus which will give rise to the gametes. The sequences lost from these chromosome ends are mostly satellite DNAs (Moritz & Roth, 1976), perhaps only required for chromosome pairing during meiosis. Even more dramatic instances of the same type have been discovered in certain insects. In the gall-midge *Wachtliella*, the somatic cells of males and females contain only six and eight chromosomes respectively, as against some forty in the germ-cell line. Over thirty whole chromosomes are eliminated from all somatic nuclei during cleavage, a complete chromosome set being retained only in the primordial germ cells and their progeny. Apparently these 30+ chromosomes are largely concerned with oogenesis, since this process fails completely in their absence, while all somatic functions are normal and even spermatogenesis can proceed to some extent. (This was shown by ligaturing embryos in such a way that *all* nuclei undergo chromosome loss; Geyer-Duszynska, 1966).

Specific DNA rearrangements occur during the differentiation of antibody-producing cells (B lymphocytes) in the vertebrate immune system. These will be treated very briefly here, since their ramifications would justify a book in themselves! Basically each antibody is a Y-shaped molecule which contains four polypeptide chains, two light and two heavy, linked together by disulphide bridges (fig. 42A). Each light or heavy chain comprises a variable (V) and a constant (C) region. Light chains are of two types designated κ and λ; in both the variable region of about 110 amino-acids spans the N-terminal half of the molecule, while the constant region (of similar length) represents the C-terminal half. Heavy chains are similar in structure, except that the constant region is much larger – some 300–400 amino-acids long. Different antibody classes (1gA, 1gD, 1gM, 1gE or 1gG) are distinguished by their different heavy-chain constant regions, designated α, δ, μ, ϵ and γ respectively (the last of these being further subdivided into $\gamma1$, $\gamma2a$, $\gamma2b$ and $\gamma3$). As the name implies, constant-region sequences are the same in all antibody molecules belonging to a given class. Variable regions, by contrast, vary widely in sequence between different antibodies within the same class. These V regions, as one might expect, are those principally involved in antigen binding. An enormous range of foreign proteins (antigens) entering the bloodstream can elicit production of highly specific antibodies by clones of B lymphocytes. How then is the requisite antibody diversity generated at the molecular level?

The number of different V-region coding sequences in the genome is large; about 300 for the mouse κ light chains, for instance. However, these V_κ genes encode only the first 95 amino-acids of the κ chain, whose variable region is 108 amino-acids long. The remaining 13 amino-acids which link this fragment onto the C_κ sequence are in fact encoded by a separate DNA sequence known as the J (joining) segment. About four different J-coding sequences occur in the mouse genome, while the C_κ gene is unique. In embryonic DNA, the V_κ genes are clustered together at a site distant from the J segment cluster, which in turn is separated from the single C_κ gene by a long intron sequence (fig. 42*B*); all three elements are located on the same chromosome. During the differentiation of κ chain-producing cells, this chromosome is rearranged by somatic recombination, so that one of the V_κ genes comes to lie immediately adjacent to one of the J sequences (Weigert *et al.*, 1980). The sequences which originally separated these two sites are deleted from the expressed chromosome (being transferred presumably to its homologue by unequal crossing over). The site of DNA exchange between the end of the V_κ gene and the start of the J segment is also variable, in that four possible base triplets can result at the junction, encoding three different amino acids. Thus the junction point itself becomes a source of heterogeneity in the variable region, over and above the 1200 combinatorial possibilities arising from 300 V_κ genes and 4 J segments. Overall, some 3000 different κ chain genes can be generated by this rearrangement. Once a V_κ gene has been positioned next to a J segment, transcription can be initiated from a leader sequence 5' to the V_κ gene through to a termination site at the 3' end of the C_κ gene (fig. 42*B*). Transcript sequences representing the long intron, together with any excess J-segment transcripts, are removed from the precursor RNA by splicing (§3.9). This generates a messenger which contains one V_κ sequence, one J sequence (that immediately following the junction point) and the one C_κ sequence (see fig. 42*B*). A similar mechanism of DNA rearrangement is involved in λ light chain production (Bernard *et al.*, 1978).

The situation is somewhat more complex in the case of heavy chain expression. Here the variable region involves three distinct DNA sequences (Early *et al.*, 1980a), i.e. one of the V_H genes (again present in numerous copies) plus two separate junctional regions designated D and J_H (each present in several genomic versions). The J_H cluster is again separated by an intron from the C_H region. Thus two somatic recombination events are required in order to juxtapose a V_H gene next to a D segment and this in turn adjacent to a J_H segment. V_H, D and J_H sequences are all clustered separately in embryonic DNA, and are brought together by somatic recombination only during B lymphocyte differentiation (fig. 42C). In humans, several different C_H genes are available, linked in the order 5′-μ-δ-γ3-γ1-γ2b-γ2a-ϵ-α-3′ on the same stretch of chromosome. Early in B lymphocyte differentiation, only μ heavy chains are expressed (1gM). Later, irreversible class switching may occur, e.g. to one of the 1gG variants. When this happens, those sequences lying between the 5′ end of the C_H cluster and the expressed gene (say γ2b) become deleted. A later class switch (say to 1gA) would delete the γ2b, γ2a and ϵ genes, so removing all the C_H genes preceding α (Cory *et al.*, 1980). In this way, the same V_H/D/J_H unit can be combined successively with different types of C_H chain. In some cases a single V_H/D/J_H unit can be expressed simultaneously in combination with two types of C_H gene (e.g. μ and δ). This probably arises from alternative splicing of long transcripts spanning much or all of the C_H gene cluster (as well as the rearranged V_H, D and J_H segments), without DNA deletion.

Each C_H gene is itself split by introns, usually into a long exon (encoding the CH1 domain), a short 'hinge' exon (H) and two further long exons (encoding the CH2 and CH3 domains; Sakano *et al.*, 1979). However, the μ gene is split into four long exons (Gough *et al.*, 1980), and its transcripts are subject to alternative processing pathways (see §3.9). The presence of an LTR-like transcriptional enhancer squence (§3.4) in the intron between J_H and the C_H gene cluster may explain why no V_H gene is expressed until one has undergone rearrangement (Gillies *et al.*, 1983; Banerji *et al.*, 1983). In the case of κ light chain gene, such an enhancer is located over 2.6 kbp downstream from the J region, within the large J-C_κ intron (see Queen & Baltimore, 1983; Picard & Schaffner, 1984). However, no such enhancer sequence is present within the re-arranged λ light chain gene-region, unless perhaps located at some distance from it. A brief summary of these events is given in fig. 42 (see also reviews by Molgaard, 1980, and by Marcu & Cooper, 1982). Overall, at least 5000 possible heavy-chain variants are encoded by the V_H, D, J_H and C_H genes. When multiplied by the available range of κ or λ light chain variants (see above), this gives well over 10^7 possible antibody molecules, encoded by a total of 10^3 or fewer gene sites.

A Domains of an antibody molecule

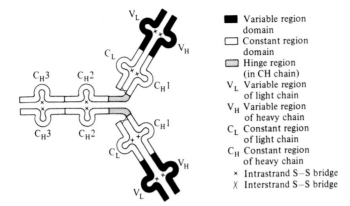

- ■ Variable region domain
- ☐ Constant region domain
- ▨ Hinge region (in CH chain)
- V_L Variable region of light chain
- V_H Variable region of heavy chain
- C_L Constant region of light chain
- C_H Constant region of heavy chain
- × Intrastrand S–S bridge
- Ⅹ Interstrand S–S bridge

B Rearrangement pathway for κ light chain genes

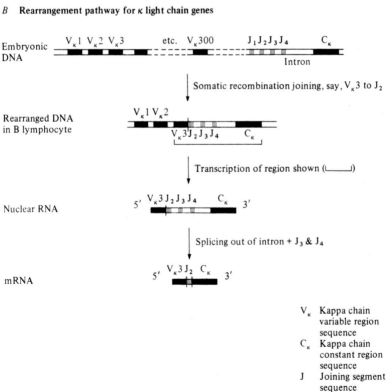

V_κ Kappa chain variable region sequence
C_κ Kappa chain constant region sequence
J Joining segment sequence

Fig. 42 Immunoglobulin gene rearrangements.

Fig. 42 (cont).

C **Rearrangement pathway for heavy chain genes (μ only)**

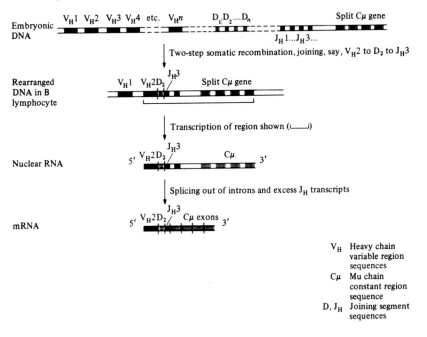

V_H Heavy chain variable region sequences

$C\mu$ Mu chain constant region sequence

D, J_H Joining segment sequences

D **Irreversible heavy-chain class switching in humans** (split structure of C_H genes omitted for clarity)

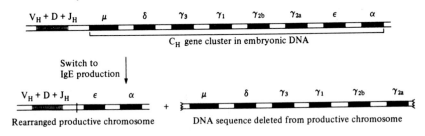

4.2.3 Nuclear equivalence

Cases of gene amplification or deletion seem limited to a few developmental systems *in vivo* (as discussed above). For most organisms, the nuclear DNA complement appears identical in all somatic tissues, as judged (crudely) by studies of DNA content and chromosomal caryotype. Of course, this is the outcome predicted from semi-conservative

DNA replication and mitosis, where each daughter cell normally receives an identical copy of the parental chromosome set. According to the differential gene expression model, most if not all nuclei in an organism contain the same set of DNA information, from which different *selections* of genes are expressed in the various cell types. Classical arguments in favour of this model are summarised in the first chapter of Davidson (1976) and elsewhere; indeed, differential gene expression has been an implicit assumption underlying much of chapters 1–3 above.

If most nuclei carry a full complement of genetic information, then in principle each of them should retain the same developmental capacities as the original zygote nucleus, which initiates normal embryogenesis. This is the concept of *nuclear equivalence* or *totipotency*. In plants, individual cells from any tissue can be induced to regenerate a complete plant under appropriate culture conditions (Steward *et al.*, 1964). Thus cells in each plant tissue contain all the DNA information necessary to generate every other cell type; they are clearly totipotent. In higher animals, this regenerative ability is much more limited, and cell-type interconversions occur only among a limited range of tissues, e.g. in the vertebrate eye system. Moreover, regeneration in animals is usually restricted to embryonic systems, or else (in adults) involves reservoirs of undifferentiated cells. Perhaps the only exception on both counts is Wolffian regeneration, where a new lens is derived from fully differentiated dorsal iris cells in adult newts (reviewed by Yamada, 1977). On this basis, animal cells do not appear totipotent; other approaches are needed in order to judge the equivalence of their nuclei.

One such approach involves nuclear transplantation into enucleated eggs (usually unfertilised). This method was pioneered in amphibian systems (e.g. Briggs & King, 1957; Gurdon, 1962; Gurdon & Uehlinger, 1966; Gurdon & Laskey, 1970), and subsequently extended to insects (Illmensee, 1972) and mammals (Illmensee & Hoppe, 1981). An unfertilised egg is first enucleated by removing its own nucleus (or a fertilised egg by removing both pronuclei), and is then micro-injected with a single diploid nucleus from a somatic tissue cell. This nucleus initiates development of an embryo which, in genetic terms, will be a clonal 'copy' of the donor individual from which the somatic nucleus was taken. At least in some cases, partial or complete embryogenesis can ensue, and a fully functional adult may eventually develop. However, several interpretative difficulties arise, mainly concerning the low success rate of these experiments. In practice, complete development is obtained far more frequently from eggs injected with early embryonic donor nuclei; successful transplants involving late embryonic or larval donor nuclei are much rarer. This could mean that non-expressed DNA regions become permanently inactivated or deleted during differentiation, so that most of the later nuclei will be unable to initiate a full programme of development in the injected eggs (Briggs & King, 1957). Those few late nuclei which do give complete development might originate from a minority population of undifferentiated 'stem cells' in the donor tissue under study. None of the tissues used successfully to date is completely immune to this criticism.

However, an alternative explanation for these results has been forwarded by Gurdon and co-workers. Most differentiated tissue cells divide slowly if at all, whereas embryonic cells divide rapidly, particularly during early cleavage. Thus a differentiated cell nucleus will probably find it difficult to make the sudden transition from mitotic quiescence to rapid cleavage divisions following transplantation. This could well explain the various chromosomal abnormalities observed during mitosis in nuclear transplant embryos, usually leading to developmental arrest. Moreover, the success rate in these transplant experiments can be dramatically improved by starting with nuclei from blastula embryos which were themselves initiated by injecting differentiated cell nuclei into unfertilised eggs. Through several such cycles, the descendants of a differentiated cell nucleus become habituated to rapid embryonic division rates, and are then able to participate in normal development without such frequent mitotic abnormalities. Even so, the overall success rate remains low, although the sheer technical difficulty of these experiments provides a plausible explanation.

Those differentiated cell nuclei which do give rise to complete organisms are clearly totipotent; that is to say, they contain all the genetic information necessary for full development, and no essential DNA has been lost or permanently inactivated during their differentiation in the donor organism.

One final point should be made here. A culture of skin cells from a *Xenopus* tadpole will never give rise to a complete toad, yet a single nucleus from one of these cells can sometimes do so following nuclear transplantation into an unfertilised egg (Gurdon & Laskey, 1970). This points to a key role for egg cytoplasm in *eliciting* the full programme of development from a totipotent nucleus. Similar conclusions can be drawn from Spemann's 1928 hair-loop experiment on early newt embryos. If a fertilised newt egg is constricted symmetrically with a hair loop so that one half contains the nucleus, then development proceeds normally in the nucleated half, while the anucleate half remains uncleaved. Around the 16- or 32-cell stage, one of the nuclei (now smaller than initially) can escape through the narrow cytoplasmic bridge connecting the two halves.

On entering the anucleate part, it can there initiate fully normal development of a second embryo. However, at this 16- or 32-cell stage no single blastomere will give rise to a complete embryo when allowed to develop in isolation. Again, the nuclei are equivalent and totipotent, but a substantial proportion of the original egg cytoplasm is necessary to elicit complete development. Oocyte cytoplasmic reserves are the topic of section 4.4.

4.3 RNA populations during development

The unfertilised egg is provided with vast stocks of RNA. Apart from ribosomes and tRNAs, these include a wide range of messenger RNAs and repetitive sequence transcripts. After fertilisation, many of these RNA species are used to direct or regulate key processes in early embryogenesis. In so doing, they exert *maternal* control over the initial stages of development (see §4.5), since the RNA sequences responsible were encoded by the mother's genes and accumulated in the egg cell during oogenesis (§4.4). As a background to this, leading on from the theme of differential gene expression (§4.2), the present section deals with changes in overall RNA populations during sea urchin development. A recent review of this topic should be consulted for further details (Davidson *et al.*, 1982).

In 1976, Galau *et al.* performed a detailed series of hybridisation experiments with mRNA populations derived from various sea urchin tissues and embryonic stages. Two types of tracer DNA were prepared from the total single-copy fraction of genomic DNA (see §1.5), by means of several cycles of mRNA–DNA hybridisation. One of these tracers, termed *mDNA*, comprised all those unique sequences which are represented in the cytoplasmic mRNA of gastrula embryos. The other tracer, termed *null DNA*, comprised all those unique sequences which are *not* represented in gastrula mRNA. Both mDNA and null DNA tracers were labelled and hybridised (separately) with an excess of mRNA prepared from each of the embryonic stages and adult tissues (see §3.2.4). From this the authors determined the complexity of mRNA sequences hybridising with each of the two probes. As an internal check, they also hybridised each mRNA population with total single-copy DNA, and showed that this overall complexity was equal to the sum of the complexities estimated separately with mDNA and null DNA tracers. (This confirms that mDNA and null DNA are mutually exclusive subsets of the original single-copy DNA fraction). In this way, they could assess the extent of both overlap and divergence between tissue mRNA populations, with respect to the gastrula messenger set (table 4).

Table 4. *Complexity of sea urchin mRNA populations using mDNA and null DNA tracers (from Galau et al., 1976)*

Tissue source of mRNA	Complexity estimated with[a]		
	mDNA	null DNA	total unique DNA
Oocyte total RNA[b]	17×10^6	20×10^6	37×10^6
Blastula mRNA	12×10^6	15×10^6	27×10^6
Gastrula mRNA	17×10^6	0	17×10^6
Pluteus mRNA	14×10^6	0.6×10^6	14.6×10^6
Adult intestine mRNA	2.1×10^6	3.7×10^6	5.8×10^6
Adult coelomocyte mRNA	3.5×10^6	1.4×10^6	4.9×10^6
Adult tubefoot mRNA	2.7×10^6	0.4×10^6	3.1×10^6
Adult ovary mRNA	13×10^6	6.7×10^6	19.7×10^6

[a] All complexity estimates given in nucleotides of mRNA sequence.
[b] Note that total cellular RNA was used in this case.

The main conclusion which emerges from these findings is that distinct but partially overlapping gene-sets are expressed as mRNA at different developmental stages. Mature oocyte total RNA is extremely complex, but only 30% more so than blastula mRNA. As differentiation proceeds, there is a general reduction in the complexity of the messenger popula-tion, a feature most obvious in three of the adult tissues compared (i.e. intestine, coelomocyte and tubefoot mRNAs). An apparent exception is adult ovary mRNA, which gives a high complexity estimate. However, this tissue will contain many immature oocytes in which high-complexity RNAs are being synthesised in preparation for the next generation. Similar results are obtained with an mDNA tracer complementary to oocyte total RNA (complexity 37×10^6 nucleotides). Some 73% of this 'maternal sequence set' is present in mRNA from 16-cell embryos, as compared to 56% in blastula mRNA and 53% in gastrula mRNA (Hough-Evans *et al.*, 1977). Some of this overlap may represent surviving maternal RNAs, but some will arise from new transcripts of the same type, i.e. expressed by the embryo's own genes.

These overlapping but distinct mRNA populations contrast with the near-identity of blastula and pluteus nuclear RNA populations (§3.10; Kleene & Humphries, 1977). Wold *et al.* (1978) found that most blastula mRNA sequences are also expressed in the nuclear RNA of adult intestine cells, but few of them are represented in the cytoplasmic mRNA of this tissue (as shown by using an mDNA preparation complementary to blastula mRNA). However, the sheer complexity of nuclear RNA populations may conceal significant differences (§3.10). Ernst *et al.* (1979) used gastrula hnRNA to prepare a null fraction from total single-copy DNA, i.e. a population of unique sequences *not* represented in gastrula hnRNA. Adult intestine nuclear RNA reacts with some 3.6% of this null DNA tracer, but even this low proportion corresponds to a sequence complexity of $>10^7$ nucleotides. Thus sequences which remain silent in gastrula nuclei are transcribed in adult intestine nuclei. Since few of these differentially transcribed sequences are represented in adult intestine cytoplasmic mRNA, they are mostly nucleus-confined and may serve a regulatory rather than coding function (see §4.9).

Most classes of repetitive DNA are represented at low or high levels in the cytoplasmic RNA of mature oocytes (Costantini *et al.*, 1978). These repetitive RNA sequences are apparently interspersed with single-copy sequences within long transcripts which overall resemble hnRNAs more closely than mRNAs (Posakony *et al.*, 1983). Possibly the repetitive elements are located in long 5' or 3' untranslated regions or else occur internally as unprocessed intron transcripts. Such repetitive transcripts are also differentially expressed in the nuclear RNAs of gastrula embryos and of adult intestine cells (Scheller *et al.*, 1978), but are absent from their cytoplasmic mRNA populations (see also §4.9). The overall picture which emerges from these studies is consistent with a model of differentiation based on differential gene expression at the mRNA and (to some extent) hnRNA levels. However, neither amplification nor deletion of some DNA sequences would necessarily result in a different pattern.

4.4 Oogenesis

The cytoplasm of the unfertilised egg (mature oocyte) is a storehouse of developmental information (Raven, 1961). If this aspect were likened to a library or data bank, then other functions of egg cytoplasm would include a machinery depot, a food warehouse, a builder's yard and a power station! Typical oocytes contain vast stocks of the following: (i) maternal mRNAs and repetitive sequence transcripts; (ii) ribosomes, tRNAs and other protein synthetic machinery; (iii) food reserves, principally in the form of yolk (also glycogen, lipids etc.); (iv) proteins required in quantity during cleavage, including histones, microtubule proteins, DNA and RNA polymerases; and (v) mitochondria. All of these reserves are utilised during early embryogenesis, and in the case of yolk the whole course of pre-hatching development depends upon egg supplies (except of course in mammals). The other functions are taken over by the embryo's own genes at different stages during embryogenesis (§4.5 below), resulting in a progressive change from maternal to embryonic genome control.

Table 5. *Comparison of oocyte reserves with those of an average somatic cell in Xenopus*

	Diameter (μm)	Ribosomes (pg)	5S + tRNAs (pg)	Nuclear DNA (pg)	rDNA (pg)	Mitochondrial DNA (pg)	Yolk (% dry weight)
Oocyte	1500	4×10^6	5×10^4	12(4C)	30(amplified)	3×10^3	45%
Somatic cell	10–50	18	3	6(2C)	Undetectable	0.06	Absent (except in female liver)

Table 5 gives some indication of the vast storage capacity of a *Xenopus* oocyte as compared with a typical somatic cell. Although large, the mature oocyte in *Xenopus* is by no means as massive as that in birds, where the entire egg 'yolk' represents a single gigantic oocyte, nearly all of which is composed of yolk. Mammalian eggs are much smaller and less yolky, since the embryo no longer depends on internal food reserves after implantation (the nutrients required for later development being supplied via the maternal placenta). In other animal groups, the size and nutritional capacity of the egg reflect the time taken for that organism to reach hatching, after which it can utilise external food sources.

4.4.1 Meroistic versus panoistic oogenesis

In most animals oogenesis follows a *panoistic* pattern. That is, the oocyte develops as an individual cell, and therefore relies in large measure on its own synthetic capacity. As we shall see below, some materials (including yolk precursor proteins) are supplied from external sources. A coat of follicle cells surrounds the growing oocyte, forming close desmosome contacts with it; these do not, however, permit free passage of macromolecules into the egg. In the mollusc *Limnaea*, these contact regions become imprinted in the structure of the oocyte. Six subcortical patches of cytoplasm – distinguished by their staining properties from the surrounding yolk – are found in an asymmetrical band around the egg, marking points of close contact with the six follicle cells. Interestingly, this cytoplasmic organisation is found only in eggs after shedding, and is not apparent in oocytes still within the ovary (Raven, 1963). Presumably the pattern imprinted by the follicle-cell contacts remains implicit until shedding, when cytoplasmic reorganisation renders it explicit (see also §4.6).

However, nucleic acid synthesis depends on the oocyte nucleus itself; thus the accumulation of large RNA stocks probably necessitates the lengthy growth period characteristic of oogenesis in panoistic systems (months or years in many vertebrates). Moreover, such specialised adaptations as ribosomal gene amplification (§4.2.1) and lampbrush chromosomes (§4.4.2) are far more prevalent in panoistic oocytes, apparently as devices for vastly increasing the rate of RNA production.

A different solution to the problem has evolved in many holometabolous insects (including *Drosophila*), whose pattern of oogenesis is termed *meroistic*. Here the oogonial stem cell undergoes a series of three or four incomplete divisions, resulting in a group of 7 or 15 *nurse cells* which remain connected to the future oocyte through open cytoplasmic

A **Ring canal system in *Hyalophora***
(diagrammatic; after King & Aggarwal, 1965)

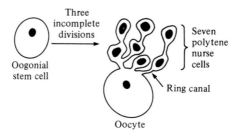

B **Incorporation of labelled cytidine in *Musca***
oogenesis (diagrammatic; after Bier, 1963)

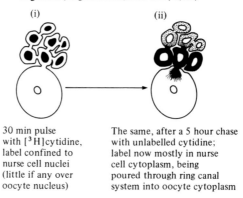

(i)	(ii)
30 min pulse with [³H]cytidine, label confined to nurse cell nuclei (little if any over oocyte nucleus)	The same, after a 5 hour chase with unlabelled cytidine; label now mostly in nurse cell cytoplasm, being poured through ring canal system into oocyte cytoplasm

Fig. 43 Meroistic oogenesis.

channels known as *ring canals* (fig. 43). Each of the nurse cell nuclei becomes polytene (see §4.2.1), which greatly magnifies the number of DNA duplexes engaged in RNA synthesis at active sites. The synthetic products of these nurse cells (both RNA and protein) are passed through the ring-canal system and accumulated in the growing oocyte (fig. 43), which consequently grows at their expense and may eventually engulf them. Throughout all this, the oocyte nucleus remains tetraploid (it is still in meiotic prophase) and seems relatively inactive in RNA synthesis. Thus large cytoplasmic reserves can be accumulated rapidly in meroistic oocytes, without the need for an extended maturation period.

4.4.2 Accumulation of oocyte reserves

(a) *Yolk.* Yolk platelets are assembled in modified mitochondria in the oocyte cytoplasm. Each platelet is principally composed of a phosphoprotein called phosvitin (MW33–34 000) and two lipoproteins called lipovitellins I and II (MWs 110–120 000 and 30–34 000). These elements are organised in a hexagonal lattice structure to form the crystalline array characteristic of yolk platelets. The yolk proteins are not synthesised in the oocyte, but instead are processed from longer precursor proteins supplied exogenously. The precursors, known as vitellogenins (MWs about 170 000) are actually synthesised in the liver, transported in dimeric form to the ovary via the bloodstream, and are taken up into the growing oocytes by pinocytosis. Typically a single vitellogenin molecule is cleaved to yield one copy each of phosvitin and lipovitellin I and II units for assembly into platelets (see also §6.4). Thus the massive stores of yolk required in the oocytes of non-mammalian species (particularly lower vertebrates) originate from vitellogenin mRNAs and proteins synthesised in the liver under hormonal control (§6.4).

(b) *Protein synthetic machinery.* Ribosomes are also stored in vast numbers (table 5) in oocyte cytoplasm. In many panoistic systems (though not in sea urchins), the 18S + 5.8S + 28S genes (rDNA repeat, see §1.8) become amplified in the young oocyte, facilitating rapid synthesis of ribosomal RNA during later growth (§4.2.1). In meroistic oocytes, a large fraction of the RNA transferred from nurse cells to the oocyte is known to be ribosomal (Klug *et al.*, 1970). The 5S rRNA genes, by contrast, are not specifically amplified in oocytes. In some species a large reserve of genomic 5S genes is pressed into service only during oogenesis. For instance, in *Xenopus laevis* some 24 000 oocyte-type 5S genes (§1.8) are transcribed very actively by pol III in young oocytes. This large burst of 5S RNA synthesis precedes the onset of active ribosomal RNA transcription (from amplified rDNA copies), though some new production of 5S RNA occurs through the rest of oogenesis. The 5S transcripts are stored in the oocyte cytoplasm as stable 7S storage particles complexed with TFIIIA (see §3.4.3) and later as complete ribosomes, so that products from the early burst of synthesis are retained during the rest of oogenesis. tRNA production follows a similar schedule to that for 5S synthesis, with very active transcription by pol III in young oocytes. Ribosomal proteins are also required in vast amounts during oogenesis; even so, their genes are not amplified.

(c) *Messenger RNAs.* The maternal control of early cleavage development (§4.5) implies that messenger RNAs synthesised during oogenesis can be stored in the egg and utilised in the cleaving embryo. This has been termed the 'masked messenger' hypothesis (see Spirin, 1966); 'masked' because the messenger-containing ribonucleoprotein (mRNP) particles found abundantly e.g. in sea urchin eggs are not used for protein synthesis until after fertilisation. It was long thought that mRNA-associated 'masking' proteins might inhibit translation in unfertilised eggs, but this view has recently been challenged. Moon *et al.* (1982) found that mRNP particles from unfertilised eggs can act as templates for protein synthesis *in vitro*, and indeed are translated just as efficiently as protein-free mRNAs prepared from them. Presumably, 'masking' factors in prefertilisation eggs might affect ribosome function (see e.g. Metafora *et al.*, 1971) rather than the messengers as such.

Among the mRNP particles stored in unfertilised sea urchin eggs, messengers coding for the histones are prominent (Skoultchi & Gross, 1973). No detectable size changes occur in any of the five major histone mRNAs between the unfertilised egg stage where they are not translated, and early cleavage stages where they are engaged in active histone synthesis (Lifton & Kedes, 1976).

The origin of maternal messenger RNA remains somewhat controversial. Some of it may arise from lampbrush chromosome transcripts (see d below), but at least in *Xenopus* this is not the major source. Rosbash & Ford (1974) measured the amount and coding specificity of poly(A)$^+$ mRNA during *Xenopus* oogenesis. Coding specificity was estimated by means of *in vitro* translation followed by protein separation on polyacrylamide gels, in order to determine the array of major proteins encoded by the messenger population. (As we have noted in §3.10 above, this identifies only the more prevalent mRNA species present, and tells us nothing about the rare mRNAs). During most of oocyte development, there is little change in either the total amount of cytoplasmic poly(A)$^+$ mRNAs, or the range of major proteins encoded thereby. Most importantly, the poly(A)$^+$ mRNA population seems to be synthesised and established in young oocytes, prior to the period of lampbrush chromosome activity (d below). Subsequently these general results have been confirmed for a large number of individual poly(A)$^+$ mRNAs by the use of cloned probes (Golden *et al.*, 1980); in all cases, non-mitochondrial messengers are synthesized and accumulated before the lampbrush phase of oogenesis. As we have seen above, 5S rRNA is stored stably through later oogenesis following its early burst of synthesis, and the same may well be true for poly(A)$^+$ mRNAs.

It remains possible, however, that small amounts of rare class mRNAs could be processed from lampbrush chromosome transcripts and added to a preexisting stockpile of moderately prevalent and abundant messengers. This might underlie the apparently contrasting result of Hough-Evans *et al.* (1979), who measured the overall sequence complexity of RNA populations (see §4.3 above) during sea urchin oogenesis. According to these authors, the RNA population of immature (previtellogenic) oocytes contains only half the range of RNA sequences present in mature (vitellogenic) oocytes. This implies that much of the RNA complexity characteristic of the unfertilised egg is added during the last few weeks of oocyte development, i.e. when lampbrush chromosomes are active.

(d) *Lampbrush RNAs.* During the long diplotene phase of meiotic prophase (often lasting weeks, months or even years), the paired chromosomes of the tetraploid oocyte nucleus become extremely decondensed. Characteristically, the duplex DNA becomes looped out from the chromosomal axis at many sites. Both the number of loops and their sizes vary between animal groups; in the newt *Triturus* each haploid chromosome set contains some 5000 loops (fig. 44*B*), each ranging in length from 50 to 200 μm (Callan, 1963), whereas in sea urchins the loops are much fewer and smaller. Organisms using the meroistic pattern of oogenesis show little if any development of lampbrush chromosome structures in their oocyte nuclei.

Apparently each loop represents a 'domain' of chromosomal DNA (see §§1.10, 2.4) held in place at its base by the chromomere matrix (more condensed). Two DNA duplexes (chromatids) lie side by side at this stage in meiosis; these are closely associated in the chromomeres and chromosomal filament (fig. 44*A*), but not in the loops themselves, which instead form symmetrically disposed pairs, each loop representing a single DNA duplex.

These DNA loops are the sites of intensive RNA synthesis, as shown for example by autoradiography after labelling with radioactive uridine. Sometimes the entire loop appears to be transcribed, but in other cases clearly delimited transcription units can be resolved by EM spreading techniques (fig. 44*C*). Individual loops may contain two more transcription units separated by non-transcribed spacer DNA, and sometimes these lie in opposite orientations (Angelier & Lacroix, 1975). The nascent RNA transcripts become complexed with a variety of nuclear proteins, some of which are specific to the transcripts from only a few defined loops (Scott & Sommerville, 1974).

The size and complexity of these lampbrush loop transcripts suggests that they are of hnRNA type. It must be emphasised that transcription rates during the lampbrush phase are extremely rapid, perhaps 100-fold faster than in normal somatic nuclei. This has been shown by direct measurements of uridine incorporation rate, and is also implied by the dense packing of polymerase molecules and their RNA products along the length of transcribed regions (fig. 44*C*). This rapid transcription rate, combined with the large number of loops, the sheer length of many

A

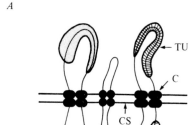

C   Condensed chromomere region at base of loop
CS   Less condensed chromosomal filament
     (two DNA duplexes side by side)
L   Decondensed loop of duplex DNA
TU   Transcription unit on loop, giving polarised
     accumulation of RNA products and
     associated proteins (as RNP)

Fig. 44*A* Semidiagrammatic representation of lampbrush chromosome loops in *Triturus* (after Callan, 1963).

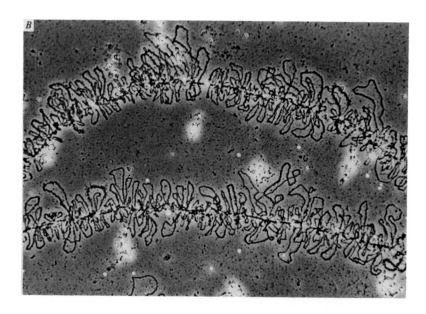

Fig. 44*B* Parts of two homologous chromosomes showing numerous lampbrush loops, from an oocyte of the newt *Notophthalmus*. Viewed under phase contrast in a centrifuged preparation. Photograph reproduced by kind permission of Prof. J. G. Gall (Carnegie Institution, Baltimore). Magnification ×434.

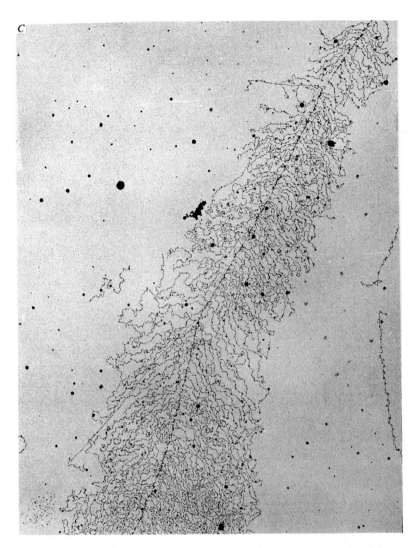

Fig. 44*C* Part of a large transcription unit in a newt lampbrush chromosome loop. Electron micrograph of a spread preparation showing central DNA axis and lateral branches representing primary transcripts. Magnification ×19 000. Photograph kindly supplied by Prof. O. L. Miller Jr. (University of Virginia, Charlottesville) and reprinted with permission from the editor of *Acta Endocrinologica*. From O. L. Miller Jr. & A. H. Bakken (1972) *Acta Endocrinol. Suppl.* **168**, 155–73.

transcription units (up to 100 kbp), and the long duration of the lampbrush phase, together imply that enormous amounts of lampbrush RNA must be synthesised. Although some of these transcripts may turn over rapidly (as with hnRNA in somatic nuclei; §3.7), large quantities of lampbrush RNA will nevertheless accumulate in the oocyte. Some of this may contribute to the stored population of maternal messenger RNAs (c above), but much or even most of it may be used for other purposes during early embryonic development.

A large fraction of lampbrush RNA consists of transcripts from repetitive DNA sequences (Costantini *et al.*, 1978; see §4.9), and even satellite DNAs are transcribed during the lampbrush phase (e.g. Varley *et al.*, 1980). In one specific instance, Diaz *et al.* (1981) have shown that long stretches of satellite DNA sequences separating the histone gene clusters are transcribed from lampbrush chromosome loops in the newt *Notophthalmus*. This probably represents 'readthrough' transcription beyond the termination signals normally observed in somatic tissues. Moreover, both strands of the satellite DNA are transcribed, though this probably results from one of the histone genes (H2B) being present in the opposite orientation to the other four (H4, H2A, H3 and H1) within each cluster (Stephenson *et al.*, 1981). Analysis of repetitive-sequence-containing RNAs from sea urchin oocytes and early embryos suggests that the repetitive elements occur interspersed with single-copy sequences, a pattern characteristic of hnRNA but not mRNA. The repetitive sequences themselves contain translational stop codons in all reading frames, suggesting that they are not expressed as protein, but may rather occur in long 5' or 3' untranslated regions, or as unprocessed intron

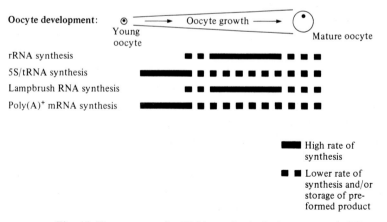

Fig. 45 Time courses for RNA synthesis during oogenesis (*Xenopus*).

transcripts (Posakony *et al.*, 1983). These transcripts are of maternal origin, but their function in the early embryo remains unclear (see Davidson *et al.*, 1982).

A summary of the time courses for synthesis of different RNA classes during oogenesis is given in fig. 45, based on the data reported for *Xenopus*.

4.5 Fertilisation and cleavage

4.5.1 Fertilisation

In this process two haploid pronuclei, from sperm and egg respectively, fuse to form the diploid zygote nucleus whose mitotic descendants will produce a complete new organism. Apart from this genetic contribution, the sperm cell adds virtually nothing in terms of cytoplasm to the huge unfertilised egg. It does, however, serve a key triggering function to initiate embryonic development. For instance, protein synthesis is activated massively following fertilisation in many invertebrate systems. At least in sea urchins, this involves a decrease in ribosome transit time, so that mRNA chains of the same length are translated over twice as fast in the zygote as in the unfertilised egg (Brandis & Raff, 1978). Most of the increase in protein synthesis, however, results from rapid assembly of new polysomes using preexisting maternal mRNAs and ribosomes. In vertebrate systems such as *Xenopus*, a corresponding increase in the rate of protein synthesis is induced hormonally prior to fertilisation, during final maturation of the oocyte. Sperm entry may also trigger a radical reorganisation of the egg cytoplasm, resulting in several distinct regions or *ooplasms*. Examples of this include the subequatorial band of red pigment granules established in zygotes of the sea urchin *Paracentrotus*, and five distinct cytoplasmic sectors formed in fertilised eggs of the sea squirt *Styela* (see §4.6). However, a similar regionalisation of the oocyte may be apparent before fertilisation in other systems.

Indeed, the whole process of fertilisation can be mimicked in many cases by parthenogenetic activation. Thus *Xenopus* or sea-urchin unfertilised eggs can initiate cleavage after pricking with a rusty needle or exposure to abnormal pH or ionic conditions. In such cases, the cleavage nuclei are of course haploid, and development usually ceases by the time of gastrulation. A few cases of natural parthenogenesis are also known, giving complete development in the absence of fertilisation. Examples include some insects (e.g. aphids) and rotifers during certain seasons of the year, and a remarkable population of all-female parthenogenetic lizards found in the Caucasus!

4.5.2 Cleavage

This first stage of embryonic development is characterised by very rapid cell division in most animal groups (with mammals as the main exception). Basically, cleavage subdivides the huge zygote cell into many smaller *blastomeres*, with a corresponding increase in the volume-ratio of nucleus to cytoplasm (approaching average somatic values in the blastula). During this process, any distinctive ooplasms in the zygote will become *partitioned* among particular groups of cells (see §4.6). In most animal groups the blastula is a hollow sphere of cells, the internal cavity being termed the blastocoele.

The rapid mitotic rate of cleaving embryos creates a high demand for several specific proteins, among them histones (needed to associate with the newly replicated DNA to form chromatin), tubulin (for mitotic spindle formation), and DNA and RNA polymerases. The activities of DNA and RNA polymerases do not increase during cleavage in either *Xenopus* or sea urchin embryos, and the demand for these enzymes seems to be met mainly from maternal stocks accumulated during oogenesis (Fansler & Loeb, 1969; Benbow *et al.*, 1975; Roeder & Rutter, 1970; Roeder, 1974). Tubulins also exist in large pre-synthesised stores in the oocyte, both in *Xenopus* (Pestell, 1975) and in sea urchins (Raff *et al.*, 1971); in addition, these proteins are newly synthesised from maternal mRNA templates during early development (Raff *et al.*, 1971). A similar dual origin is found for the histones during cleavage. In *Xenopus*, large stocks of maternal histone proteins are found in the egg (Woodland & Adamson, 1977), but this supply is supplemented by *de novo* synthesis, first from maternal mRNAs and later from new messengers encoded by the embryo's own histone genes (Woodland *et al.*, 1979). The oocyte stocks of histone H1A protein are mostly accumulated during early oogenesis in *Xenopus* (van Dongen *et al.*, 1983). In sea urchins, significant stores of histone proteins are *not* supplied by the egg (Bentinnen & Comb, 1971); instead the requirement is met by new histone synthesis from maternal messengers (Skoultchi & Gross, 1973), which are again supplemented by histone mRNAs of embryonic origin (see also §4.7 below).

A high rate of protein synthesis is also required in order to sustain the rapid production of histones etc. in cleaving embryos. The maternal stock of ribosomes is so vast that little new synthesis of rRNA occurs during cleavage. In *Xenopus*, o_{nu}/o_{nu} homozygotes with little rDNA of their own (§1.8) can survive to the swimming tadpole stage using only these maternal ribosomes; even in normal embryos, rRNA synthesis remains at a relatively low level until this stage. Apparently a few of the rDNA repeat units are active during embryogenesis, while the majority remain repressed. Similarly in sea urchins, ribosome production is not fully activated until the larval feeding-pluteus stage. In mammals, however, active rRNA synthesis is initiated during cleavage, probably because the maternal ribosome stocks are much smaller. Activation of tRNA synthesis occurs during the late cleavage or gastrula stages in most animals.

By contrast, hnRNA synthesis begins very early during cleavage in most types of animal embryo, usually between the 2- and 16-cell stages. At least in some cases, messenger RNAs derived from these transcripts are utilised during normal cleavage development (e.g. new histone mRNAs in sea urchin and *Xenopus*, as mentioned above). Figure 46 shows a schematic summary of the time courses for synthesis of different RNA classes from the embryonic genome.

Although many embryo-coded messengers are transcribed (and some even translated) during cleavage, these do not appear essential for the cleavage process itself, which proceeds largely under *maternal* control. The drug actinomycin D has been used to block transcription during early embryonic development in a wide range of animal species. In most cases, cleavage occurs normally, but gastrulation is blocked. At face value, this would suggest that the mRNAs synthesised during cleavage are a *preparation* for later developmental stages, rather than for immediate use. Unfortunately, actinomycin has several toxic side effects, does not effectively block synthesis of all types of RNA if given at low dose levels, and may also fail to penetrate the embryo properly. Even where appropriate contols have been performed, the results of actinomycin experiments remain ambiguous (see chapter 2 in Davidson, 1976).

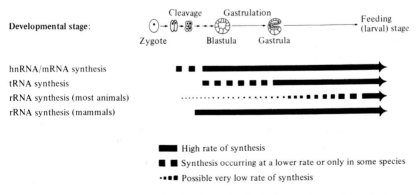

Fig. 46 Time courses for RNA synthesis during early development.

However, a variety of other experiments point to the same conclusion; namely, that cleavage processes are predominantly under maternal control, and can continue in the absence of any contribution from the embryonic genome. The most dramatic evidence for this is provided by the partial blastulae obtained when enucleated eggs from sea urchins (Harvey, 1936) or *Rana* (Briggs *et al.*, 1951) are activated parthenogenetically. Cell walls subdivide the egg cytoplasm and amphiaster figures are formed within the 'cells', despite the total absence of chromosomes. Many specific features of cleavage development are also under direct maternal control. One example from classical embryology concerns the number of primary mesenchyme cells produced in late sea urchin blastulae. In cases where viable hybrids can form between two sea urchin species, each characterised by different numbers of these cells, the hybrid embryos always follow the maternal pattern (table 6; Driesch, 1898). Another example is the direction of spiral displacement between different tiers of cleavage blastomeres in gastropod molluscs, a feature which is later reflected in left- or right-handed coiling of the shell. In some normally right-handed (dextral) species such as *Limnaea*, left-handed (sinistral) coiling is inherited as a simple one-gene recessive character; however, phenotypic expression of the coiling genotype always lags one generation behind the standard Mendelian pattern (Boycott *et al.*, 1930). Presumably, it is the *maternal* genes, acting through factors accumulated in the egg cytoplasm, which determine the direction of shell coiling in all of the offspring (fig. 47).

The duration of maternal control during early development varies considerably between different animal groups, and also according to the parameter examined (some functions being taken over by embryo genes sooner than others; see e.g. fig. 46). In sea urchins, new embryo-coded variants of histone H1 appear as early as the 8-cell stage, in addition to H1 histones derived from existing maternal mRNAs (see §4.7 below). In *Xenopus*, paternal H1 histone genes are activated in the midblastula stage, while stored maternal H1 messengers disappear in the early

Table 6. *Numbers of primary mesenchyme cells in a sea urchin interspecies hybrid (after Driesch, 1898)*

♀ species	♂ species	Average number of primary mesenchyme cells
Sphaerechinus × *Sphaerechinus*		33 ± 4
Echinus × *Echinus*		55 ± 4
Sphaerechinus × *Echinus*		35 ± 5

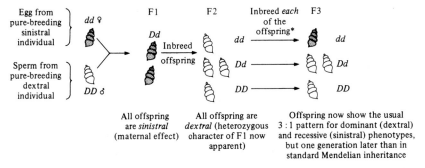

Fig. 47 Maternal inheritance of shell coiling pattern in the gastropod *Limnaea peregra*: D, dominant allele for dextral coiling; d, recessive allele for sinistral coiling. *N.B. These molluscs are hermaphrodite.

gastrula. This was shown ingeniously by Woodland *et al.*(1979), using 'interspecies hybrids' between *X. laevis* enucleated eggs (containing maternal cytoplasmic reserves but no chromosomal DNA) and *X. borealis* sperm (providing paternal H1 genes of a different type). In both *Xenopus* and sea urchins, the new embryo transcripts first appear while maternal messengers of the same general type are still available. This probably explains why cleavage can proceed in embryos treated with actinomycin D (to block new transcription), although gastrulation is prevented.

In mammals the onset of embryo genome control occurs during the early cleavage divisions, with many genes being activated at the late 2-cell stage. A case in point is that of β_2 microglobulin in mouse development; a mutant form of this protein encoded by the paternal gene (hence absent from the maternal reserves) can be detected in embryos as early as the late 2-cell stage (Sawicki *et al.*, 1981). A rather more involved experiment leading to similar conclusions was performed by Monk & Harper (1978); in early mouse embryos, the level of an X-linked enzyme, hypoxanthine phosphoribosyl transferase (HPRT), shows maternal genome influence (XX or XO) only up to the 8-cell stage, after which the embryo's own genotype (several possible permutations) assumes control over HPRT levels. Maternal HPRT mRNA is gradually replaced by embryo-coded messengers during early cleavage (Harper & Monk, 1983).

However, this difference between mammals and other animal groups as regards the timing of embryo genome activity, may be more apparent than real. Strictly it applies only to developmental stage and not to 'real' time, because mammalian embryos cleave much more slowly than those of other animals. By the time a mammalian zygote has cleaved once or twice, a *Xenopus* or sea urchin embryo will have reached the blastula stage or beyond. Thus maternal influence may persist for a similar number of hours in both situations, though not for the same number of cleavage divisions.

4.6 Mosaic versus regulative development

4.6.1 Mosaic systems and cytoplasmic determinants

One problem which we have skirted so far is how groups of embryonic cells are directed to follow particular developmental pathways. We have seen that maternal influences extend well into cleavage development (and in some cases beyond). If this maternal information were unequally distributed between different parts of the egg cytoplasm, it would become partitioned during cleavage into discrete groups of blastomeres, and could then direct each group towards a particular fate. In essence, this is the pattern of *mosaic* development found in many invertebrate groups, including molluscs and ascideans (sea squirts). In such systems the zygote cytoplasm is highly organised, and is often subdivided into regions distinguished by different pigmentation or yolk density.

A classic case of this type was described in the sea squirt *Styela* by Conklin (1905). Here, sperm entry triggers a process of cytoplasmic streaming in the egg, from which five distinct sectors emerge: clear cytoplasm, pale grey and dark grey yolky regions, plus pale and dark yellow crescents on one side of the egg. By the 64-cell stage (late cleavage) each of these cytoplasmic sectors has been segregated into a different group of blastomeres. More importantly, each of these blastomere types has a distinct developmental fate. Thus cells containing the clear cytoplasm develop into ectoderm, those with pale grey material into notochord, those with dark grey yolky cytoplasm into endoderm, those with pale yellow crescent material into coelomic mesoderm, and those with dark yellow crescent material into the larval tail muscles. Moreover, these cell fates are irrevocably fixed. If the cells containing one type of egg cytoplasm are destroyed (blastomere deletion), then the embryo develops without the structures normally formed by those cells, and cannot 'regulate' to make good the loss. If instead such blastomeres are removed and allowed to develop in isolation, they form only the appropriate tissue type and no others. Thus embryos of this type develop as mosaics of cells whose future fates are largely mapped out by their inheritance of cytoplasmic substances (termed *morphogenetic determinants*) from different regions of the fertilised egg.

Of course, it is not always possible to see obvious signs of localisation in the zygote, but blastomere isolation and deletion experiments reveal a mosaic pattern of development in many other invertebrate systems. These include the small soil nematode *Caenorhabditis elegans*, for which complete and largely invariant cell lineages have been described, covering the whole of embryonic and post-embryonic development (Deppe *et al.*, 1978; Sulston *et al.*, 1980; Sulston & White, 1980; Laufer *et al.*, 1980; Sulston *et al.*, 1983). Notably this organism contains a fixed number of cells (959 somatic nuclei in the adult hermaphrodite), allowing the ancestry of each cell to be traced back to the initial zygote.

An interesting example of mosaic development is provided by certain gastropod (*Ilyanassa*) and scaphopod (*Dentalium*) molluscs. Here the egg contains regions of clear cytoplasm both at the animal pole (containing the nucleus) and at the opposite vegetal pole (polar plasm), separated by a wide band of coloured yolky material. Prior to first cleavage, the vegetal polar plasm becomes extruded as a *polar lobe* connected to the rest of the zygote by a narrow cytoplasmic bridge. During first cleavage the polar lobe is passed to only one of the two daughter blastomeres, into which it is resorbed on completion of the division. The lobe-carrying cell is designated CD, its lobeless companion AB. At second cleavage the same thing happens, so that the polar lobe is transferred to only one cell (D) out of the four (C, like A and B, is lobeless). This is shown diagrammatically in fig. 48 (see Wilson, 1904a,b).

If this polar lobe is cut off prior to either first or second cleavage, then the embryo which develops is deficient in mesodermal tissues such as muscle etc. Thus the polar lobe presumably contains mesodermal determinants, directing those cells which inherit them to develop towards a mesodermal fate. Although the pattern of polar lobe formation breaks down during later cleavage, the mesodermal determinants continue to be shunted into one particular daughter cell through several further divisions, as shown by blastomere deletion experiments in *Ilyanassa*

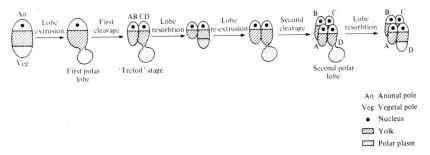

An Animal pole
Veg Vegetal pole
• Nucleus
▨ Yolk
▢ Polar plasm

Fig. 48 Early cleavage in *Dentalium* or *Ilyanassa* (diagrammatic).

(Clement, 1962). After the sixth cleavage, a single cell designated 4d contains the mesodermal determinants, and from it are descended the muscle cells of the later embryo; hence 4d is also known as the primary mesentoblast cell.

However, the actual cytoplasmic contents of the polar lobe during the first or second cleavage divisions seem relatively unimportant to its function. If *Dentalium* zygotes are centrifuged gently, the yolky material can be displaced to the vegetal pole. Nevertheless, a polar lobe still forms at its normal vegetal site, even though it is now filled with yolk instead of clear polar plasm. Such an embryo is able to develop normally, and moreover removal of the polar lobe prior to first or second cleavage still results in a mesoderm-deficient embryo (Verdonk, 1968). So mesodermal determinants must still be localised in the polar lobe region, despite its altered cytoplasmic contents after centrifugation. One likely explanation is that these determinants occur embedded in the *cortex* of the polar lobe region, rather than free in the polar plasm. The egg cortex is a layer of gelated cytoplasm lying immediately under the plasma membrane, and is not displaced by mild centrifugation. In other systems, including *Styela*, mild centrifugation of the zygote tends to disrupt development, implying that the morphogenetic determinants are more easily displaced.

Cytoplasmic localisation plays a key role in germ-cell determination in many animal groups. Clear examples of this are afforded by those organisms where the somatic nuclei undergo chromosome diminution during cleavage (e.g. *Ascaris*, *Wachtliella*; see §4.2.2). Only one or a few nuclei escape this process, namely those destined to found the germ-cell line. The primordial germ-cell nuclei are always located at one pole of the embryo (posterior pole in insects), in a region of cytoplasm distinguished by granular inclusions termed polar granules. If nuclei are not permitted to enter this region (e.g. by ligaturing the embryo; Geyer-Duszynska, 1966) or if the same region is cauterised by narrow-beam UV irradiation before any nuclei have migrated into it, then germ cells fail to develop. In cases of chromosome diminution, this means that *no* nuclei retain their full chromosomal complement. UV irradiation of the polar region has the same effect in many organisms which do not undergo chromosome diminution; i.e. no germ cells are formed, and the adults developing from such embryos are sterile (no gametes). The effects of UV-cauterisation can be reversed by subsequently microinjecting cytoplasm from the corresponding polar region of an unirradiated egg; germ cells will then develop normally from any nuclei migrating into the injected region. This cytoplasm must therefore include germ-cell determinants, possibly associated with the characteristic polar granules.

Many classic experiments on germ-cell determination have used insect systems – mainly in species such as *Drosophila* which do not show chromosome diminution; the main advantage of these organisms is their peculiar pattern of *superficial* cleavage (fig. 49).

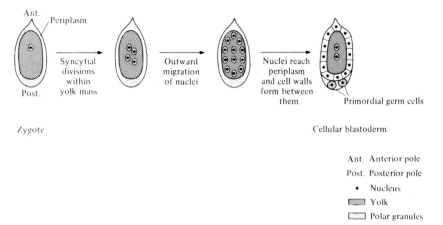

Fig. 49 Insect cleavage (diagrammatic).

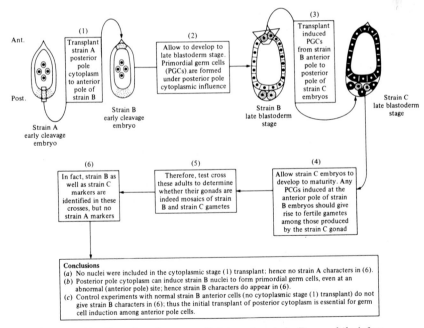

Fig. 50. Cytoplasmic germ cell determinants in *Drosophila* (after Illmensee & Mahowald, 1974).

Thus cauterisation and microinjection procedures involving the posterior pole region do not affect the early cleavage nuclei if performed prior to their outward migration. Illmensee & Mahowald in 1974 provided an elegant demonstration that cytoplasm from the posterior pole region can induce germ cell formation even when injected at an unusual (anterior pole) site. Their experiment, involving three strains of *Drosophila* carrying different genetic markers, is summarised in fig. 50.

As to the chemical nature of morphogenetic determinants, this remains almost totally obscure. Polar granules have been isolated from the posterior pole cytoplasm of early *Drosophila* embryos, and a basic protein of MW 95 000 was identified as their major constituent (Waring *et al.*, 1978). However, there are no reports that this purified material can induce germ-cell formation, for example in UV-irradiated embryos. A subcellular fraction prepared from *Drosophila* eggs by Ueda & Okada (1982) can apparently induce the formation of 'pole' cells but not germ cells when injected into the posterior (but not anterior) pole of UV-irradiated eggs. This complexity suggests that several determinants may act in concert in the normal posterior pole region.

In the Dipteran insect *Smittia*, double-abdomen embryos (no head or thoracic structures) can be induced by UV irradiation of the anterior pole prior to blastoderm formation, while centrifugation during cleavage induces double-head embryos (no abdominal structures). Jäckle & Kalthoff (1980, 1981) have investigated the spectrum of proteins being synthesised in anterior and posterior fragments of normal, double-head and double-abdomen embryos. They found one protein consistently present in all fragments destined to form head structures (anterior indicator protein), and another present in all fragments destined to form abdomen (posterior indicator protein). However, there is as yet no proof that either of these proteins is a morphogenetic determinant. Several different maternal-effect mutations in *Drosophila* result in dorsalised embryos lacking ventral structures. Some of these mutant phenotypes can be rescued by injecting poly(A)$^+$ RNA from wild-type embryos (Anderson & Nusslein-Volhard, 1984). Of the ten maternal loci affecting dorsal/ventral polarity, only the *Toll* gene seems likely to encode a true morphogenetic determinant (Anderson, reported in North, 1984).

4.6.2 Regulative systems

In embryology, the term *regulation* means that an embryonic system is able to make good deficiencies caused by removing some part of that system. Whereas early cleavage blastomeres from mosaic embryos show restricted developmental fates, those from *regulative* embryos are able to develop a complete set of tissues. In the starfish *Asterina*, all 8 blastomeres from an 8-cell embryo can regulate to form complete $1/8$-sized larvae (Dan-Sohkawa & Satoh, 1978). In sea urchins, regulative ability persists up to the 4-cell stage but is later lost.

Nevertheless, there is good evidence in sea urchin systems for cytoplasmic localisation, which may sometimes be visibly apparent (e.g. the subequatorial pigment band in *Paracentrotus*; §4.5.1). From about the 8-cell stage onwards, sea urchin blastomeres show restricted developmental fates (fig. 51). Thus when tiers of blastomeres from 16-cell embryos are separated and grown in isolation, the three cell types develop differently (fig. 51). The mesomeres (a ring of 8 cells at the animal pole) form *animalised* embryos, consisting mainly of ectodermal derivatives but deficient in endoderm. The macromeres (four larger cells in the vegetal half of the egg) form *vegetalised* embryos composed mainly of endoderm (gut), but lacking a normal complement of ectoderm. The micromeres (four very small cells from the vegetal pole) generally do not develop in isolation, but have been shown to form mesodermal elements (skeletal spicules) when cultured *en masse* (Okazaki, 1975).

However, these distinct developmental fates are already latent in the fertilised sea urchin egg. If a zygote is bisected horizontally between the animal and vegetal poles, then in cases where the animal half receives the nucleus, an animalised (mainly ectoderm) embryo develops; in cases where the vegetal half receives the nucleus, the embryo which develops is vegetalised (deficient in ectoderm). Similar effects can be obtained by treating whole embryos with chemical agents; lithium chloride causes embryos to become vegetalised (inhibiting ectoderm formation), while sodium thiocyanate causes them to become animalised (inhibiting endoderm/mesoderm formation).

These and other experiments led to the double-gradient hypothesis of Runnström (1928) and Hörstadius (1928). In essence this postulates an ectoderm-promoting influence centred on the animal pole and decreasing towards the vegetal pole, together with an endoderm/mesoderm-promoting influence centred on the vegetal pole and declining towards the animal pole (fig. 51). Normal development depends on a balance between these two gradients. For instance, combinations of macromeres (fairly high vegetal influence) with mesomeres (high animal influence) will regulate, i.e. develop into complete embryos, despite the absence of micromere material. Perhaps more surprisingly, complete embryos are also formed from combinations of mesomeres with micromeres (highest level of vegetal influence). In this case the resulting embryos are dwarf, since they lack the macromere material representing nearly half the volume of a

16-cell embryo. However, such micromere/mesomere combinations do *not* develop into an endoderm-deficient embryo composed mainly of mesoderm and ectoderm, as would be expected in a typical mosaic system. Intercellular interactions are clearly required in order to redistribute the gradient influences across the remaining embryonic material, such that a normal complement of endoderm also forms. This is an important feature of regulative systems and pattern formation in general (see Wolpert, 1969). However, it is a topic beyond the scope of the present text.

As mentioned above, individual sea urchin blastomeres from both 2- and 4-cell stages can regulate to form complete embryos. Yet the double gradient system of morphogenetic influences appears to be established before first cleavage, as revealed by the zygote bisection experiments (fig. 51). This apparent paradox is simply resolved; the first two cleavage divisions are in a vertical plane, and do not cut across the axis of either gradient influence (see fig. 51). In other words, each 2- or 4-cell-stage blastomere contains the same balance of animal and vegetal influences as does the original zygote. Only when the third cleavage cuts across the gradient systems in a horizontal plane do the blastomeres lose their equivalence. Each individual blastomere from the 8-cell stage will contain a preponderance of either the animal or vegetal influence, and is thus no longer able to give rise to a normal balance of embryonic tissues.

This relationship between cleavage planes and the segregation of morphogenetic determinants can be used to explain the apparently regulative or mosaic nature of most types of embryo (see chapter 7 in Davidson, 1976). In mosaic embryos, even the first cleavage results in an unequal segregation of these determinants (obvious in the case of *Dentalium*; fig. 48), so that neither daughter cell has the same balance of developmental potentialities as the zygote. In regulative embryos, on the other hand, each early cleavage blastomere receives a complete and balanced set of determinants, and it is only during later cleavage divisions that some determinants are apportioned preferentially into one group of cells rather than another. At this stage, the cells will cease to be equivalent in their developmental capacities. Even so, such cells do not follow a completely mosaic fate thereafter, as discussed above. One simple way to account for this supposes that the morphogenetic factors are diffusible between cells in regulative embryos, but not in mosaic embryos. In fact, many intergradings between these patterns are possible, and most mosaic embryos show evidence of intercellular communication during their later development (e.g. shell-gland induction in molluscan embryos). Further subtleties are introduced by the fact that cytoplasmic localisation is

A **Normal cleavage pattern**

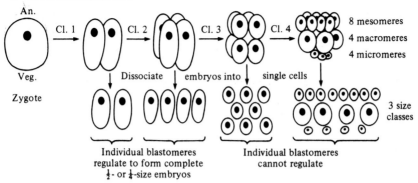

Individual blastomeres
regulate to form complete
½- or ¼-size embryos

Individual blastomeres
cannot regulate

B **Animalised embryos**

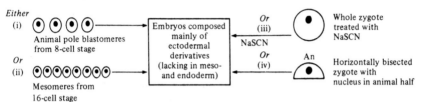

C **Vegetalised embryos**

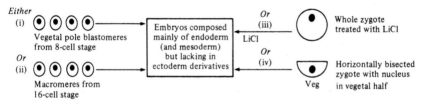

D **Regulation and the double-gradient system** (16-cell stage)

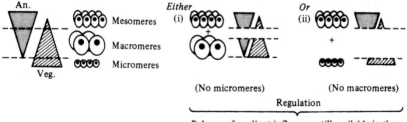

(No micromeres)   (No macromeres)

Regulation

Balance of gradient influences still available in these
combinations, hence a complete embryo develops
(albeit undersized in the case of (ii))

▨ 'Animal' influence (ectoderm-promoting)

▨ 'Vegetal' influence (endo/mesoderm promoting)

Fig. 51 Early sea urchin development.

sometimes established *progressively* during the early cleavage divisions, rather than preformed in the egg or zygote. A clear example of progressive localisation is provided by the segregation of two sets of determinants during cleavage in the ctenophore *Mnemiopsis* (Freeman & Reynolds, 1973; Freeman, 1976).

A major exception to these generalisations is found in mammalian embryos, where cell position exerts an overriding influence on future development. Thus cells located on the outside of the 16–32 cell *morula* become extra-embryonic trophoblast derivatives (e.g. placenta) following cavitation, while those on the inside develop into the embryo proper (inner cell mass). Up to the cavitation stage, cells transplanted from the inside to the outside or *vice versa* will develop in accordance with their new location. After cavitation, however, cells from the inner cell mass cannot transform into trophoblast if exposed on the outer surface, and trophoblast cells injected into the interior cannot turn into inner cell mass derivatives. There is as yet no convincing evidence for cytoplasmic localisation in mammalian systems, and as we have seen above (§4.5.2), maternal control is generally lost during the early cleavage stages. This is presumably related to the much slower division rates found during early mammalian development.

4.7 Establishment of molecular differences between early blastomeres

When do cells or cell-groups first become distinct from each other during early development? In the preceding sections we have looked at this question in terms of restricted developmental capacities; to what extent can these be related to differences at the molecular level?

In the mollusc *Ilyanassa*, the spectrum of proteins being synthesised in AB-cell derivatives shows consistent differences from that in CD-cell derivatives, following blastomere separation at the 2-cell stage (Donohoo & Kafatos, 1973). Although suggestive, these results do not show whether qualitatively different maternal messenger sets are present initially in the lobeless AB and lobe-carrying CD cells, nor whether special mRNAs might be localised in the polar lobe. In *Styela* embryos, *in situ* hybridisation with an actin DNA probe reveals that actin mRNA is differentially distributed, with very little present in those cytoplasmic regions destined to form endoderm, but a high proportion present in the yellow crescents destined to form mesoderm (Jeffery *et al.*, 1983). These yellow crescent regions (mesoplasm) contain only 5% of the total poly(A)$^+$ mRNA of the embryo, but nearly half of the total actin mRNA sequences. Although actin is a characteristic product of mesodermal cells, it appears that the actin mRNAs present in *Styela* embryos encode predominantly cytoplasmic rather than muscle-specific isoforms of actin (see §1.9).

For the remainder of this section we will focus on the 16-cell stage of sea urchin development (fig. 51). Here, the three blastomere types already show restricted fates (§4.6); they can also be isolated in bulk from synchronous 16-cell embryos, taking advantage of their different cell sizes.

Quantitative but no qualitative differences are detectable when protein synthetic profiles are compared between meso-, macro- and micromeres (Senger & Gross, 1978). Actinomycin treatment suppresses these quantitative differences, so that the protein synthesis patterns become indistinguishable between the three cell types. This suggests that all three contain similar populations of maternal messenger RNAs, since actinomycin blocks transcription of the embryo's own genes. In other words, prevalent maternal mRNAs (those identifiable at the protein level) are distributed homogeneously throughout the 16-cell embryo, and presumably also in the zygote. The fact that quantitative differences in protein synthesis arise only when the embryo genome is transcribed, does not necessarily imply that different amounts of messenger RNA are synthesised in the nuclei of mesomeres, macromeres and micromeres. Even if all sixteen nuclei synthesised identical amounts and types of mRNA, quantitative differences would still be observed because of the cell-size difference at the 16-cell stage, resulting in different ratios of maternal to embryonic mRNAs (table 7). This shows how quantitative differences could arise between blastomeres without any requirement for differential transcription or for uneven distribution of maternal messengers, but merely as a consequence of unequal cleavage divisions leading to differences in blastomere size. It does not of course explain why cleavage should follow this particular pattern.

In fact, a detailed analysis of the RNA populations in meso-, macro- and micromeres suggests that neither maternal RNAs nor new transcripts are homogeneously distributed. Rodgers & Gross reported in 1978 that micromeres lack a proportion of the RNA species present in both meso- and macromeres. This was shown by preparing an 'egg$^+$ DNA' tracer, comprising all those single-copy sequences complementary to total oocyte RNA (the maternal sequence set; see §4.3 above). This tracer was hybridised with an excess of total cellular RNA prepared respectively from isolated meso-, macro- and micromeres, as well as from whole 16-cell embryos. Egg$^+$ DNA reacted to the same extent (around 90%) with total RNA from mesomeres, macromeres and 16-cell embryos, but

Table 7. *Senger & Gross (1978) model to explain quantitative differences in protein synthesis between meso-, macro- and micromeres*

	Cell type	Cytoplasmic maternal mRNA[a]	Embryonic mRNA[b]	Overall messenger population	
A. In normal embryos	Mesomere	MMMM	NN	4M : 2N	Different ratios of M
	Macromere	MMMMMM	NN	6M : 2N	and N
	Micromere	MM	NN	2M : 2N	proteins will be translated from each
B. In actino-mycin-treated embryos	Mesomere	MMMM	—	All M	Only M pro-
	Macromere	MMMMMM	—	All M	teins will be
	Micromere	MM	—	All M	translated from each

[a] Since prevalent mRNAs are distributed homogeneously, their contribution in each cell type should reflect cytoplasmic volume (macromere > mesomere > micromere).
[b] This contribution is assumed to be constant for all 16 nuclei, in the simplest case.

to a markedly lower extent (67–80%) with total RNA from micromeres. This means that the maternal sequence set (egg[+] DNA sequences) is less fully represented in the micromeres, i.e. they contain a more limited range of RNA species than do the other cell types in 16-cell embryos. Since almost identical results were obtained in actinomycin-treated embryos, this difference cannot be ascribed to new transcription, but probably reflects an inhomogeneous distribution of maternal RNAs in the zygote.

This intriguing result prompted Tufaro & Brandhorst (1979) to reexamine in greater detail the spectrum of proteins being synthesised in mesomeres, macromeres and micromeres. Despite a high-resolution 2-dimensional gel technique able to resolve up to 1000 different proteins, they could detect only quantitative differences between the three cell types. Specifically, none of the proteins synthesised in meso- and macromeres were absent from the micromere pattern. These protein data at first sight seem to contradict the inhomogeneous RNA distribution inferred above.

The apparent paradox here was eventually resolved by Ernst *et al.* in 1980. They found that both meso- and macromeres contain rare-class maternal mRNA species which are *not* present on polysomes (i.e. are not being translated into protein). By contrast, the micromeres do not contain any detectable reserves of non-translated mRNAs. Similarly, both meso- and macromeres contain high-complexity nuclear RNA species (transcribed from the embryo's own genes), whereas micromeres do not. Thus the overall complexity of micromere total RNA is indeed less than that of meso- and macromere total RNAs (as found by Rodgers & Gross). However, the RNA sequences lacking in micromeres are not polysomal (i.e. not translated) in meso- and macromeres, hence the qualitative identity in protein synthetic profiles (as found by Tufaro & Brandhorst).

The limited RNA complexity in micromeres may underlie their restricted developmental potential, but does not explain the similarly restricted potentials of meso- and macromeres. In a sense this is hardly surprising; micromeres represent a very small fraction of the total volume of 16-cell embryo cytoplasm, so that local differences would be expected to show up clearly. A similar local difference confined, say to the animal pole, could be swamped by the much greater volume of mesomere cytoplasm. Localised accumulations of poly(A)$^+$ RNA have been reported in *Xenopus* zygotes and oocytes (Capco & Jeffery, 1981, 1982), though their developmental significance is not known.

The establishment of spatial differences within the early embryo is presumably accompanied by temporal changes. One clear-cut example of this has been reported in sea urchins by Senger *et al.* (1978). The normal form of histone H1 expressed during early cleavage is termed $H1_m$, being synthesised at least in part from maternal mRNA templates. Messenger RNA encoding a new H1 variant designated $H1_x$ is actively transcribed from embryonic genes during the late 8-cell stage. This results in a rapidly rising ratio of $H1_x$ to $H1_m$ messengers, which is reflected in their respective rates of protein synthesis, and in the amounts of these two H1 histones incorporated into new chromatin. In late 16-cell embryos, however, the ratio of $H1_x$ to $H1_m$ falls back to its previous low level, and this again applies at the mRNA, protein and chromatin levels. Consistent differences appear at the 16-cell stage between meso-, macro- and micromeres, in terms of the $H1_x : H1_m$ ratios incorporated into their chromatin. How this relates to the developmental divergence of these three cell types is uncertain; as mentioned in section 2.2, replacement of histone subtypes may lead to altered nucleosome stability (Simpson, 1981). This in turn could affect the transcriptional capacity and/or rate of replication of the chromatin.

As indicated above, mass cultures of isolated micromeres can differentiate into primary-mesenchyme derivatives such as skeletal spicules (Okazaki, 1975). Harkey & Whiteley (1983) have shown that the pattern of protein synthesis changes markedly in these cells at around the time of hatching in whole embryos, and also in mass cultures of isolated micromeres. This change involves decreased synthesis of early proteins (characteristic of 16-cell stage micromeres) and enhanced synthesis of late (differentiated) proteins. Thus the programme of primary-mesenchyme differentiation is already mapped out (*determined*; see next section) but is not yet expressed in the micromeres of 16-cell embryos.

4.8 Differentiation

4.8.1 *Determination versus differentiation*

Several examples of differentiating systems will be discussed in part 2 of this book, so only general features are considered below. As discussed previously (§§3.10, 4.2), the specialised cell types in an adult animal differ fundamentally in their utilisation of 'luxury' genes, but overlap to some extent in their 'housekeeping' functions. Both quantitative and qualitative changes may alter the pattern of gene expression during the course of differentiation in a given cell type. However, the appearance of tissue-specific *markers* (e.g. enzymes or other proteins) in significant amounts is usually taken as diagnostic, and for practical purposes will serve as a definition of differentiation.

Cells often commit themselves irreversibly to a particular fate well before they express any of its differentiated characteristics. This is the embryological phenomenon known as *determination* or *commitment* (the two terms tend to be used interchangeably). Perhaps the most striking example of determination in advance of overt differentiation is provided by the imaginal discs involved in insect metamorphosis, a topic to be discussed in chapter 7. Other examples are mentioned below.

Myoblast cells do not express differentiated muscle characteristics while actively dividing in culture; they will only proceed to accumulate muscle-specific mRNAs and proteins after passing through a terminal division prior to fusion into syncitial myotubes (John *et al.*, 1977). The dividing myoblasts are determined in that they cannot develop into other cell types, but differentiation is delayed until the postmitotic stage and subsequent fusion. Several transformed cell-lines show a similar distinction between division and differentiation in culture. Thus dividing populations of neuroblastoma cells or MEL cells (erythroid precursors infected with Friend virus) can be induced to cease division and to differentiate as neurones or red blood cells respectively, following treatment with polar agents such as dimethyl sulphoxide (DMSO). Presumably these cell lines are arrested in the 'determined precursor' state, with differentiation blocked until triggered by the inducing agent (which seems fairly non-specific and is certainly non-physiological). Again, only a single differentiation pathway is open to such cells, so they are already determined towards that fate even though they do not express its characteristic products while dividing.

Variant lines of mouse erythroleukaemia (MEL) cells have been isolated, in which the inducer-mediated processes of erythroid differentiation and commitment to the terminal cell division can be dissociated from each other. A similar distinction can also be made by using different types of inducing agent; e.g. haemin induces erythroid differentiation without commitment to the terminal cell division, whereas DMSO induces both processes. Shen *et al.* (1983) have recently shown that commitment to the terminal cell division is associated with a decrease in the level of one particular nuclear protein of MW 53 000. This protein does not decrease during erythroid differentiation in the absence of commitment to the terminal cell division.

In molecular terms, determination might involve the *selection* of particular gene sets, making them 'available' for expression, e.g. as DNase I-sensitive smooth-fibre chromatin, perhaps flanked by DNase I-hypersensitive and/or demethylated sites (§2.6). However, significant transcription from these genes can only ensue when some further signal (e.g. hormonal or inductive) is received to initiate differentiation.

4.8.2 Induction

In many specific instances, differentiation depends upon a short-range *inductive* interaction between tissues of different embryonic origin, which are brought into close proximity by the movement of cell masses during gastrulation and later development. For instance, formation of the vertebrate eye lens from head ectoderm requires an inductive influence (inducer) from the optic cup, which is an outgrowth of the anterior neural tube (fig. 52A). Indeed, neural development itself depends upon induction(s) from the prechordal plate and notochord material; during gastrulation these tissues invaginate under the ectoderm of the future neural plate (primary embryonic induction; fig. 52B). The region of cells showing *competence*, i.e. ability to respond to a particular inductive influence, is often larger than that part of it which finally differentiates along the pathway specified; this allows some degree of regulation by the induction system. Moreover, both competence and inducer-availability are usually limited to a fairly brief period during development. Thus hierarchies of

A **Eye lens induction** (Diagrammatic transverse sections of anterior neural tube)

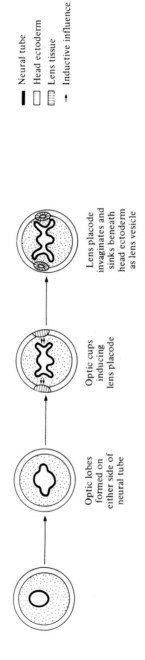

Optic lobes formed on either side of neural tube

Optic cups inducing lens placode

Lens placode invaginates and sinks beneath head ectoderm as lens vesicle

▮ Neural tube
▢ Head ectoderm
▨ Lens tissue
↑ Inductive influence

B **Primary embryonic induction** (Diagrammatic longitudinal sections of *Xenopus* embryo)

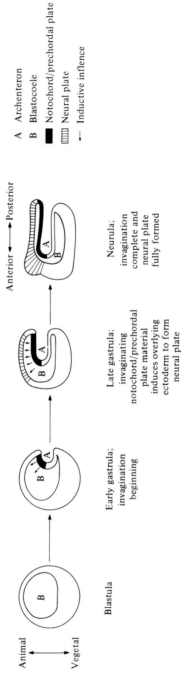

Animal ↕ Vegetal

Anterior ⟷ Posterior

Blastula

Early gastrula; invagination beginning

Late gastrula; invaginating notochord/prechordal plate material induces overlying ectoderm to form neural plate

Neurula; invagination complete and neural plate fully formed

A Archenteron
B Blastocoele
▮ Notochord/prechordal plate
▨ Neural plate
→ Inductive inflence

C **Epithelio-mesenchymal induction** (transfilter experiment)

E Epithelial tissue

M Mesenchymal tissue

Permeable (e.g. Millipore) filter

Reciprocal inductive influences (not always apparent in E ---→ M direction

Specific glandular differentiation from epithelial component

Possible specific differentiation from mesenchymal component

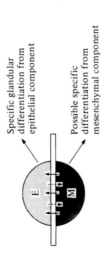

Fig. 52 Examples of tissue induction.

successive inductions can be built up, as commonly observed during organ formation (e.g. in the eye).

In many organ systems, the two interacting tissues are a *mesenchyme* (mesodermal) and an *epithelium* (ecto- or endodermal) respectively. Either one or both components respond to induction with a characteristic pattern of differentiation; usually the epithelium develops specific glandular structures, while the mesenchyme may or may not show obvious signs of change. No differentiation occurs if an impermeable barrier is inserted between the two tissues; however, if a permeable barrier is used (such as a Millipore filter) then induction proceeds normally. This implies that diffusible molecules rather than actual cell-cell contacts are involved, although very fine cytoplasmic processes might possibly make contact through the filter pores (see e.g. Smith & Thorogood, 1983). An *in vitro* induction system has been devised by Clifford Grobstein, in which an epithelial and a mesenchymal tissue are co-cultured on either side of a Millipore filter (fig. 52C), allowing their reciprocal interactions to be studied. The specificity of these inductions can be gauged by comparing homologous with heterologous tissue combinations (where, respectively, the two components are derived from the same or from different organ sources). In some cases the epithelium will respond only to homologous mesenchyme, but in others a variety of mesenchymes will elicit the same type of epithelial differentiation. This method has identified cases where two or more distinct inductions are required for the full differentiation of a single tissue type (e.g. in liver and kidney; see Wolff, 1968).

The identity of these inducing agents remains as elusive as that of the morphogenetic determinants discussed in sections 4.5 and 4.6. There are many reports claiming to have isolated proteins or other macromolecular factors which show specific inductive effects, especially in primary embryonic induction. Unfortunately, these effects can be mimicked by a variety of wholly non-specific agents, ranging from altered pH and ionic conditions to organic dyes such as methylene blue! A plausible explanation for this supposes that one or more *potential* differentiation programmes might already be mapped out (implicit) in the responding cells, but can only become expressed (explicit) when triggerd by appropriate inducing agents. As we have seen in the case of 'determined precursor' cell-lines, an effective inducer *in vitro* may be a molecule never encountered *in vivo*.

4.8.3 Developmental and hormonal controls

Developmental control is a general term used in systems where the pattern of gene expression changes with time as part of an overall differentiation programme. Such controls can operate at several levels; the quantitative ratios of certain gene products may alter because of changes in the rate of mRNA production and/or translation, alternative RNA processing pathways may be used (§3.9), or related genes may be expressed sequentially (see chapter 5). Some recent data suggests that sequences responding to developmental signals can lie within the transcribed portion of a gene, as well as in its 5′-flanking regions. This has been shown e.g. by fusing regions of the human β globin gene, either 5′ or 3′ from the translational initiation site, to complementary parts of heterologous genes; when such constructs are introduced into MEL cells, human β globin transcription is inducible by DMSO in both cases (Wright *et al.*, 1984).

Hormones afford another means for inducing differentiation, or for changing the pattern of gene expression in already differentiated cells. Their characteristic is long-range action, as opposed to the short-range effects of inducing agents. In section 2.2 we saw how hormones binding at the target cell surface could alter such parameters as histone phosphorylation within that cell, with possible consequences for cell division and altered gene expression. A less roundabout route to the same end is taken by many steroid hormones. Their mechanism of action has been worked out in several systems (see chapter 6). The essential steps involved are as follows (fig. 53):

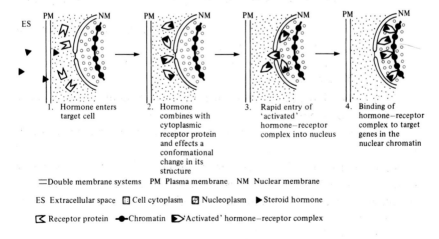

1. Hormone enters target cell
2. Hormone combines with cytoplasmic receptor protein and effects a conformational change in its structure
3. Rapid entry of 'activated' hormone–receptor complex into nucleus
4. Binding of hormone–receptor complex to target genes in the nuclear chromatin

═Double membrane systems   PM Plasma membrane   NM Nuclear membrane

ES Extracellular space   ▣ Cell cytoplasm   ▣ Nucleoplasm   ▶ Steroid hormone

◿ Receptor protein   ◆Chromatin   ▷'Activated' hormone–receptor complex

Fig. 53 Essential steps in steroid hormone action. Recently it has been shown that unoccupied steroid receptor proteins are found in the nuclei and not in the cytoplasm of target cells. This would eliminate stages 2 and 3, allowing the steroid to bind directly to receptor proteins within the nucleus.

(i) Hormone enters the cell; this probably does not require a specific uptake system, since steroids are hydrophobic molecules which pass readily through the cell membrane.

(ii) The hormone binds to a specific *receptor* protein present only in the cytoplasm of target cells. This causes a conformational change (activation) in the receptor protein molecule.

(iii) The activated hormone–receptor complex rapidly enters the nucleus.

(iv) Specific DNA sequences in the 5′-flanking regions of hormone-sensitive genes are recognized by the hormone–receptor complex, either activating or blocking transcription.

Recent evidence suggests that steroid receptor proteins may be located in the nuclei of target cells at all times, and that their apparent presence in the cytoplasm prior to hormone binding may be an artefact (see Welshons *et al.*, 1984; King & Greene, 1984). This would simplify the above scheme by combining stages (ii) and (iii). Whether hormone receptor proteins are confined to the neighbourhood of their target genes within such nuclei remains to be determined.

The example of oestrogen action in the oviduct and liver will be considered in chapter 6, and the role of ecdysone during insect metamorphosis in chapter 7.

4.9 A model for the regulation of gene expression in eucaryotes

Any such model must take account of – and if possible explain – a seemingly disparate array of facts. These include:

(a) The presence of a large fraction of repetitive sequences in eucaryotic DNA (chapter 1).

(b) Short-period interspersion of repetitive elements among longer single-copy sequences (§1.6); this arrangement seems to characterise much of the DNA in most animals (the few exceptions have very small genomes).

(c) Transcription of a substantial fraction of the genome as highly complex hnRNA (§3.7), representing both unique and repetitive sequences; these are partly interspersed as in (b).

(d) The instability (short half-life) of this hnRNA, much of which is nucleus-confined (§§3.7, 3.10).

(e) The complexity of rare-class messengers reaching the cytoplasm (see §3.10); some of these are common to several cell types, but many are expressed in a tissue-specific manner.

(f) There is a limited set of discrete pathways in development, and not an infinitude of overlapping states of differentiation; any cell will be directed (canalised) towards one or other of these basic patterns.

(g) Multiple changes in cellular gene expression can be elicited by relatively simple effectors, such as inducing agents or hormones (§4.8).

In greater or lesser measure, a rationale for all of these features is provided by the Britten–Davidson model of eucaryotic gene regulation to be discussed below. In fact, many of them could equally be explained in other ways; for instance (d) above seems consistent with what we know of split gene expression and intranuclear RNA processing (§§3.8, 3.9). However, when the hnRNAs and corresponding mRNAs from a single cell type are considered as whole populations, it becomes apparent that only *some* of the former act as precursors to the latter (§3.10).

Certain general implications of the above list are obvious. For example, (f) and (g) both suggest that gene regulation is *integrated* in some way, so that large 'batteries' of genes can be activated or shut down together. In bacteria, functional integration is achieved through coordinate transcription of genes organised into operon groups. In principle, differentiation in higher organisms could also involve huge operon-like structures. However, the distances between functionally related genes are often vast (e.g. the α and β globin genes are located on different chromosomes in higher vertebrates; chapter 5), and there is evidence from several systems that any one such gene can be regulated independently of other members of the 'battery'. This emerges from many studies of enzyme distribution between different cell types (reviewed in Truman, 1974, 1982). An enzyme (or other protein) characteristic of one pathway of differentiation sometimes appears in a different context in an unrelated cell type. For instance, the enzyme dopa-decarboxylase in *Drosophila* is involved both in neurotransmitter synthesis (catecholaminergic nerve cells) and in cuticle pigmentation and crosslinking (hypodermal cells). Moreover, the gene coding for this enzyme is single-copy, arguing against the possibility that the same gene might recur in two (or more) 'operon' batteries. What is required then, is an integrated control system capable of much greater flexibility than that provided by the procaryotic operon model.

A model for gene regulation providing such flexibility was forwarded in 1969 by Britten & Davidson. It should be emphasised that the model itself is fairly flexible, and has been adapted in various ways since its original publication to account for new findings in eucaryotic molecular biology (e.g. Davidson *et al.*, 1977; Davidson & Britten, 1979).

The Britten–Davidson model proposes five elements in a regulatory network. *Sensor* sites are DNA sequences or DNA-protein complexes which can respond to simple effector stimuli (f above). These, together with the *producer* genes coding for messenger RNAs, would feature in any model of gene control. The three distinctive elements of the model are required as links between the sensors and the producer-gene batteries whose expression they control. *Integrator* genes adjacent to the sensor would be transcribed only when appropriate signals are received at the sensor site. Their transcripts, termed *activator* RNAs, would then interact with *receptor* sites adjacent to appropriate producer genes, thereby switching on transcription of the producer RNAs. In principle this might be achieved through translation of the activator RNA into a protein with specific affinity for the receptor sequence, but Britten & Davidson (1969) favoured the simpler alternative of direct RNA–DNA interactions between activator and receptor.

The integrative function of these extra elements comes from the proposal that both integrator genes and receptor sites are *repetitive* DNA sequences. One particular integrator sequence might occur adjacent to several different sensors, so that various stimuli could activate the same gene, or set of genes. This latter possibility would arise if the same receptor sequence were found adjacent to *all* the producer genes in a particular battery. Each producer gene could have multiple receptor sites in its 5' flanking region, so that a variety of different activator RNAs would switch on its expression. Similarly, multiple integrator genes could be controlled from a single sensor. In this way it is possible for a signal received at one sensor to activate a large battery of producer genes (g above), while a different signal received at another sensor could activate a partially overlapping set of producer genes (e above).

The number of producer genes activated by particular signals would then depend on the *redundancy* (number of copies) of both integrator genes and their corresponding receptor sequences. Similar considerations could apply to the switching-off of previously expressed producer genes (negative rather than positive control). If activator RNA interacts directly with receptor DNA, then some measure of sequence complementarity between them is predicted by the model (this would not apply if protein activators were used). Simple examples of such control networks are shown schematically in fig. 54. It is easy to see how far more complex patterns of gene expression could arise through combinations of both receptor and integrator redundancy. Several specific predictions arise from this general model.

First of all, the receptor sites; if these exist, then repetitive sequences should be found in the neighbourhood (especially near the 5' ends) of most producer genes. Moreover, the *same* repetitive sequence (receptor) should recur in or close to all genes subject to a common regulatory pathway – i.e. the members of a particular gene-battery. Some evidence on this point is now available from direct sequencing studies. Closely-related sequences are indeed found near the 5′ ends of several rodent and human genes regulated by glucocorticoid hormones (Cochet *et al.*, 1982); a similar situation applies to oestrogen-regulated genes in chick, and to yeast genes coding for enzymes involved in the same biochemical pathway (reviewed in Davidson *et al.*, 1983). These common sequences are not always identical within a 'battery', and are anyway rather short (9–30 bp), which means that conventional techniques for detecting repetitive sequences would have missed them altogether. These sequences lie between 60 and 400 bp upstream from the mRNA cap site, which suggests that they are not transcribed. Whether such common sequences will be found for other gene-batteries, and whether they are of sufficient specificity to fulfil an integrative receptor function, must obviously await further work. In the case of hormone-regulated genes, common flanking sequences may well represent hormone–receptor binding sites (see §6.3).

A second point concerns the redundant integrator genes proposed by the model. Repetitive DNA is known to occur in many different sequence families (about 4000 in sea urchin), each ranging in copy number from a few tens to many thousands per haploid genome. During sea-urchin development, many of these repetititve families are transcribed in a tissue- and stage-specific manner (Scheller *et al.*, 1978); thus sequences represented abundantly at one stage may be rare or absent at another. Such transcripts, representing most of the available repetitive families, are present in mature oocyte cytoplasm as part of the maternal sequence set inherited by the embryo (Costantini *et al.*, 1978); they may well originate from lampbrush RNA during oogenesis (§4.4). As discussed earlier, these repetitive sequences do not themselves code for protein, but occur interspersed among single-copy sequences as parts of long transcripts which seem to be of hnRNA-type despite their cytoplasmic location (Posakony *et al.*, 1983). It remains to be seen whether such RNAs can be processed to typical messengers during embryogenesis, or whether they serve some other (regulatory) function in early development (see review by Davidson *et al.*, 1982). It is possible that these maternal repetitive transcripts could influence blastomere nuclei during cleavage, perhaps differentially via cytoplasmic localisation (see §§4.6 and 4.7 above). At other developmental stages, however, transcripts of these repetitive sequences are confined to the nucleus (as predicted by the model), and often both strands of the template are represented among them. All of this is consistent with – but does not prove – an integrator role for repetitive sequence families. Repetitive sequence families are also transcribed in many other organisms, including *inter alia* certain transposable elements (e.g. *copia*) in *Drosophila*; a recent brief review of this topic is given by Davidson & Posakony (1982).

This leads us on to the third major prediction, concerning the nature of activator RNAs. Davidson *et al.* (1977) have argued that much of the nucleus-confined hnRNA is in fact 'activator' RNA (or integrated regulatory transcripts, IRTs). The interspersed repetitive nature of much hnRNA (Smith *et al.*, 1974) is consistent with this proposal, as is its restriction to the nucleus. If recognition between the activator RNA and receptor site DNA involves some form of complementary base-pairing, then a minimum concentration of activator RNA will be needed within the nucleus to maintain the base-paired structure. If a single activator RNA species is required to interact with many receptor sites (fig. 54), then much higher concentrations of that activator will be necessary. This

A **Redundancy in receptor sites**

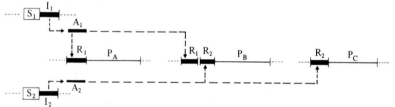

A Activator RNA   I Integrator gene   P Producer gene   R Receptor site   S Sensor site

Producer genes P_A and P_B are transcribed in response to signals received at S_1, while P_B and P_C similarly respond to S_2. Only P_B responds to both

B **Redundancy in integrator genes**

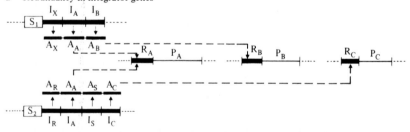

Producer gene P_A would be activated in response to signals at both S_1 and S_2, while producer genes P_B and P_C would be activated only by signals at S_1 and at S_2 respectively. Activators A_R, A_S, A_X would act on yet other genes

Fig. 54 Integrative model for gene control (simplified from Britten & Davidson, 1969).

could underlie both the differing copy-numbers observed among repetitive sequence families ('integrator' genes), and the changing levels of their transcripts ('activator' RNAs) during development. Large gene-batteries would be expected to respond only to abundant activator RNAs transcribed from highly redundant integrator genes, while small batteries could respond to less prevalent activator RNAs derived from smaller sets of integrator genes. One key prediction emerges from this: the receptor sequences characterising all producer genes in a large battery should be complementary to parts of the abundant activator transcripts in any cell-type where that battery is active. Though this point should be testable experimentally, there is as yet no direct evidence for or against it.

So far it has been implicitly assumed that activator–receptor recognition involves RNA–DNA base-pairing, so switching on transcription of producer genes adjacent to the receptors. This is not in fact a necessary feature of the model. Davidson & Britten (1979) have proposed an alternative version, in which the activator hnRNA base-pairs instead with the transcripts of producer genes. In essence, this modification suggests that most producer genes are transcribed constitutively at a fairly constant basal level in all cell types throughout development. These primary transcripts would be processed to messengers only if they were recognised by an activator RNA with appropriate complementarity, i.e. to receptor sequences included in the producer transcript. Thus formation of RNA:RNA duplexes within the nucleus would determine whether a particular producer transcript is processed to mRNA or not. This could explain why many protein-coding genes are transcribed in the nucleus, yet are not represented in the cytoplasmic mRNA (e.g. Wold *et al.*, 1978; see also §3.10). More generally, it could suggest why such a high proportion of the single-copy genome is expressed in nuclear hnRNA (§3.10). In neither case does the available evidence imply that *all* producer genes are transcribed in the nucleus, but a measure of post-transcriptional selection among transcripts is certainly required.

Finally, the modified model can offer some explanation for different mRNA abundance-classes in the cytoplasm. The efficiency of transcript processing would presumably reflect the rate of RNA duplex formation, which would in turn depend on the intranuclear concentrations of the RNA reactants (producer gene transcripts and activator RNAs). If most producer genes are transcribed constitutively, then different activator RNA levels (see above) could affect the rate of processing for producer gene transcripts. This seems adequate to account for rare and moderately prevalent mRNA production, but breaks down in the case of abundant (superprevalent) messengers. Additional controls are required in these cases, since their genes are probably transcribed much faster than the basal constitutive rate proposed for other genes. In any case, some form of transcriptional control is necessary even for the modified model, since activator RNAs would have to be transcribed differentially between different tissues.

The above discussion does not necessarily imply that any one version of the model is correct. However the basic formal system proposed by the original model has not changed, and this is really more important than the detailed superstructures built upon it. In one version or another, the model can account for many of the peculiar features of eucaryotic gene expression; – therein lies its value.

In two mammalian systems, certain classes of repetitive sequences (termed 'identifier' or ID sequences) have been shown to characterise many of the transcripts expressed in a particular tissue or developmental stage. One such identifier sequence, 82 nucleotides long, has been found in some 62% of brain transcripts, but not in other tissues (Sutcliffe *et al.*, 1982). Another such sequence characterises many transcripts which are prominent during early mouse development, and also in pluripotential transformed cells; the representation of these transcripts declines as differentiation proceeds (Murphy *et al.*, 1983). Both types of ID sequence occur dispersed at many sites in the genome, and they are both transcribed by pol II as parts of longer RNAs which also contain unique-sequence transcripts. The brain-type ID sequences are located within the introns of many genes expressed specifically in brain, but also occur elsewhere in the genome (Milner *et al.*, 1984) – suggesting that this ID sequence may be a mobile genetic element (see §1.10). There is also evidence that much shorter pol III transcripts contain these same ID sequences. It remains to be seen whether analogous identifier elements will be found to characterise the transcripts (and gene-batteries) expressed in other major cell-types. The existence of ID sequences need not necessarily imply a regulatory function *à la* Britten and Davidson, so this attractive possibility must await a direct experimental test. This might be achieved by linking an identifier DNA sequence to a heterologous gene, and then introducing the chimaeric gene into early embryos. One can then ask whether that gene becomes expressed inappropriately in a tissue whose transcripts are normally characterised by the ID sequence (e.g. would a globin gene plus brain ID sequence become expressed in brain tissue?) However, one prerequisite for this approach will be a gene-transfer system where introduced genes-plus-flanking-sequences are routinely expressed under normal developmental control (e.g. the P-element transfer system in *Drosophila*; see § 7.2).

PART 2

Specific Systems

5

Erythroid differentiation

Summary

Vertebrate erythrocytes contain large quantities of the oxygen-carrying protein haemoglobin, which is composed of four globin chains (two α-type, two β-type) each linked to a haem prosthetic group (§5.1).

In adults the main site of erythroid differentiation is the bone marrow, where a self-renewing population of stem cells gives rise not only to erythrocytes, but also to various other types of circulating blood cell (lymphocytes, granulocytes, megakaryocytes etc.). Commitment to follow one or other of these pathways occurs during the early divisions of one daughter cell derived from a stem-cell mitosis (the other daughter, on average, must remain a stem cell). Subsequently, each committed erythroid precursor cell undergoes a series of amplification divisions stimulated by the glycoprotein hormone erythropoietin, resulting in a vast number of erythrocytes from each precursor cell (§5.2).

Both the site of erythroid differentiation and the types of globin chain expressed change characteristically during human development (§5.3). Adult bone-marrow-derived erythroid cells express α, β and δ globins; foetal erythroid differentiation occurs in the liver, with the α and γ globin genes active (but not β or δ); in the embryo, the earliest erythroid cells are derived from the yolk sac and express ϵ and ζ globin chains in addition to α and γ. These switches in globin chain type result not only from the different tissue sources of embryonic, foetal and adult erythroid populations, but also from changes in the pattern of globin gene expression within preexisting erythroid cells. The changing chromatin structure of the globin gene regions during development is also briefly reviewed in section 5.3.

Features of globin gene transcription mentioned earlier (chapters 1–4) are recalled in section 5.4, and a few of the many regulatory mechanisms affecting globin expression in erythroid cells are explored. Finally, section 5.5 reviews the molecular organisation of the globin genes; in humans, these form linked clusters of α-related (α, ζ) genes on chromosome 16 and of β-related (β, δ, γ, ϵ) genes on chromosome 11. The evolution and molecular pathology of these genes is also briefly discussed.

5.1 Introduction

The aim in the second part of this book is to explore a few differentiating systems in more detail. In so doing, the order of the first part will be reversed; starting with a differentiated cell, we shall work back through its development to examine the transcription and control of cell-type-specific genes, before finally considering the organisation of those genes in the DNA. Globin genes and their expression in erythroid cells have been mentioned in many different contexts during the first four chapters; they are undoubtedly the most intensively studied genes in eucaryotes. In the account below, examples discussed earlier will not be repeated in detail, but instead a brief summary and/or appropriate section reference will be used to cover the point.

Each globin protein chain becomes complexed to an iron-containing haem molecule which acts as a prosthetic group for oxygen-binding. Two α-type and two β-type globin chains (plus attached haems) form the subunits of a tetramer called *haemoglobin* (Hb), which is characterised by cooperative allosteric oxygen-binding. This means that binding of O_2 to the haem of one subunit causes conformational changes in the other subunits which facilitate O_2-binding at the three remaining haem sites. Oxygen release in the tissues is promoted by CO_2 and H^+ there, which bind to different but allosterically linked sites in the haemoglobin subunits. The converse happens where oxygen is freely available (in the lungs or gills), with CO_2 and H^+ being released and new O_2 bound. Vertebrate erythrocytes are circulating blood cells packed full of haemoglobin; they are thus highly specialised for carrying oxygen from the lungs or gills to the tissues where it is required, and also for CO_2/H^+ transport in the reverse direction (for details, see Stryer 1981, chapter 4). Although haemoglobin is the major differentiation marker in erythroid (i.e. erythrocyte precursor) cells, a series of haem-synthesising enzymes and other proteins are also essential for proper cell function.

In the remaining sections of this chapter, we shall consider: (i) the origin and differentiation of red blood cells in adults; (ii) the switches in haemoglobin type which occur during development; (iii) the expression and control of globin genes; and (iv) their organisation in the DNA.

5.2 Differentiation of erythroid cells in adults

Erythrocytes, and indeed all other varieties of blood cell, are constantly renewed throughout life. Each type of blood cell has a characteristically limited life span, averaging 120 days for human erythrocytes, but considerably longer for some granulocytes and lymphocytes. Worn-out cells are phagocytosed by macrophages in the liver and spleen, while new cells are constantly produced from special blood-forming sites. These are located mainly in the adult bone marrow, but accessory sites include, for example, the spleen. In humans, the average rate of erythrocyte production is around two million cells per second! The source of these new cells is a dividing population of *stem cells* which remains within the bone marrow. In fact, a single type of *pluripotent* stem cell can give rise to a wide variety of blood-cell types, as indicated in fig. 55. At each stem-cell division, one of the daughter cells effectively remains a stem cell (hence this population is indefinitely self-renewing), while the other undergoes a limited series of divisions. This latter type of cell becomes committed during its early divisions to follow one of the several available pathways of blood-cell differentiation (fig. 55).

Specific glycoprotein factors are required in order to stimulate division among particular kinds of committed precursor cell. These include (i) erythropoietin for cells in the erythroid pathway (see below), and (ii) granulocyte/macrophage colony-stimulating factor (GM-CSF), which at low concentrations favours macrophage production and at high levels promotes granulocyte formation. Each committed precursor cell undergoes a series of *amplification divisions* under the appropriate glycoprotein influence; this has nothing to do with gene amplification, but is simply a means for generating large numbers of differentiated cells from a single precursor. In the case of the red blood cell lineage, high concentrations of erythropoietin stimulate the division of early erythroid precursor cells; these are known as BFU-Es (erythroid burst-forming units) because each such precursor gives rise to a large 'burst' containing many thousands of differentiated erythrocytes. This involves eleven or twelve successive cell

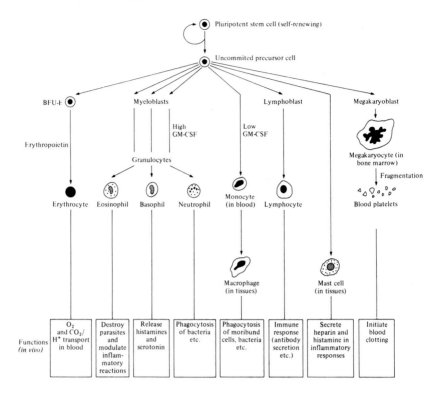

Fig. 55 Pathways of blood cell differentiation (adapted from Alberts *et al.*, 1983 and Ede, 1978).

divisions in each clone founded by a single BFU-E, as shown *in vitro* using low-density cultures of bone marrow cells. Low concentrations of erythropoietin, on the other hand, stimulate division only among later erythroid precursors; these are known as CFU-Es (erythroid colony-forming units) because each such cell can undergo only five or six further cell divisions before final differentiation. Hence a small colony of 60 or fewer erythrocytes is produced from each CFU-E at low erythropoietin levels, as compared to the large burst of many thousand red blood cells formed from each BFU-E at high erythropoietin levels (fig. 56).

There is a sequence of characteristic changes in cell ultrastructure and staining properties during the last few cell divisions in the erythroid series (fig. 57). This leads from the proerythroblast through basophilic, polychromatic and orthochromatic erythroblast stages, the last of which gives rise directly to the reticulocyte without further cell division. Terminal differentiation from reticulocyte to erythrocyte is also accomplished without cell division; indeed in mammals the nucleus is eliminated during

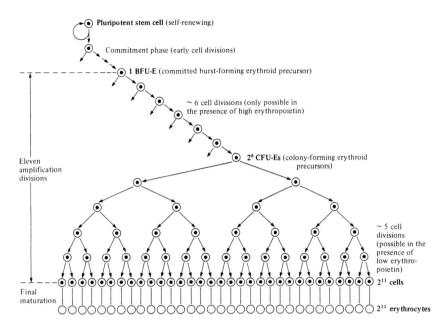

Fig. 56 Erythroid cell lineage (modified from Alberts *et al.*, 1983).

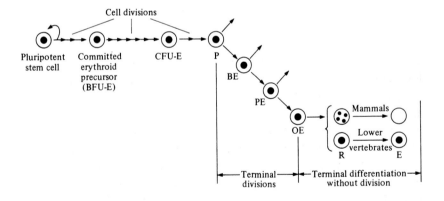

P   Proerythroblast (first signs of haemoglobin synthesis)
BE Basophilic erythroblast
PE Polychromatic erythroblast
OE Orthochromatic erythroblast
R   Reticulocyte (nuclear loss in mammals)
E   Erythrocyte (nucleated in lower vertebrates)

Fig. 57 Terminal divisions and differentiation in the erythroid lineage (modified from Ede, 1978).

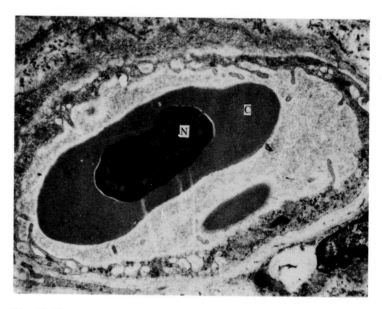

Fig. 58 Electron micrograph showing a nucleated avian erythrocyte within a blood capillary supplying the oviduct of a laying hen. Note the homogeneous cytoplasm (C), dark (inactive) nucleus (N) and flattened disc shape (seen in cross section). Magnification ×5500. By courtesy of Dr I. R. Duce (Department of Zoology, University of Nottingham).

the reticulocyte stage (fig. 57), though other vertebrates such as birds (fig. 58) have nucleated erythrocytes. At least in birds, histone H5 is involved in the inactivation of erythrocyte chromatin (§2.2). Globin mRNA and protein (differentiation markers) first appear in small amounts after the CFU-E stage (e.g. in proerythroblasts), and they are accumulated rapidly during later differentiation. However, precursors from the BFU-E stage onwards are irreversibly committed to an erythroid fate (see also §4.8). Haemoglobin synthesis continues during the reticulocyte stage, despite loss or complete inactivation of the nucleus; this implies that globin mRNA is stable and can be translated for some time after transcription of new messengers has ceased.

5.3 Haemoglobin switching

In adult humans the predominant type of haemoglobin is HbA, composed of two alpha and two beta subunits ($\alpha_2\beta_2$). Both α and β globins are produced by different but related members of the globin gene family (§1.9). In adults there is also a small percentage of HbA$_2$ (2–3%), composed of two α and two delta subunits ($\alpha_2\delta_2$). The δ chains are derived from a further member of the globin gene family closely related to β. Thus in adult bone-marrow-derived erythroid cells, the β, δ and α globin genes are all actively transcribed. In fact there are two α genes but only one β and one δ gene (see §5.5 below); nevertheless the levels of α and β production are closely coordinated so as to achieve a $1:1$ ratio. This is mediated in part by translational control mechanisms, involving for instance the haem-precursor haemin (see §5.4 below; Giglioni *et al.*, 1973; Lane *et al.*, 1974). The forty-fold disparity between β and δ levels results in large part from differences in the rate of mRNA production, but δ mRNA is also more rapidly degraded than β mRNA (Wood *et al.*, 1978) and may be intrinsically less stable.

During embryonic and foetal development, however, alternative sites of erythropoiesis (erythrocyte formation) and several further globin genes are pressed into service. In human foetuses, red blood cells are produced by the liver. The α genes are active in these foetal erythroid cells, but the β and δ genes are not. Instead, two β-related gamma genes are expressed, giving rise to foetal haemoglobin (HbF; $\alpha_2 \gamma_2$). The two γ globin genes are not identical, though their protein sequences differ by only one amino-acid (glycine in $^G\gamma$, alanine in $^A\gamma$). HbF has a higher oxygen affinity than HbA, hence oxygen can be transferred from the maternal circulation (HbA) to the foetal blood system (HbF) via the placenta.

In human embryos, two special embryonic globin genes are expressed; these are the α-type zeta (ζ) gene and the β-type epsilon (ϵ) gene; both α and γ genes are also active at this stage. This results in a series of embryonic haemoglobins, namely Hb Gower 1 ($\zeta_2 \epsilon_2$), Hb Gower 2 ($\alpha_2 \epsilon_2$) and Hb Portland ($\zeta_2 \gamma_2$). These are expressed only in early erythroid cells originating from the embryonic *yolk sac*, which is the earliest site of erythropoiesis in higher vertebrates. The ζ and ϵ genes are shut down completely in adult and foetal erythroid cells. These changes are summarised in table 8.

Similar changes in globin type and erythropoietic site are found in other mammals, and to some extent in other vertebrates. Thus chickens express embryonic types of haemoglobin only in the primitive erythroid cells derived from the extraembryonic blastoderm (blood islands) at early stages of development, but switch to adult types in the definitive erythroid cells produced in later embryos. Similarly, amphibia have distinct tadpole and adult haemoglobin types adapted to the different life styles at these two stages.

From all this, it might be thought that switches in haemoglobin type result simply from replacement of one erythroid cell population by

Table 8. *Haemoglobin gene switching in man (based on Weatherall & Clegg, 1979)*

Erythropoietic tissue	Stage	β-type genes active	α-type genes active	Globin products
Yolk sac	Embryo	ϵ and $^A\gamma/^G\gamma$	ζ and α	Hb Gower 1 $\zeta_2\epsilon_2$ Hb Gower 2 $\alpha_2\epsilon_2$ Hb Portland $\zeta_2{}^{A/G}\gamma_2$
Liver	Foetus	$^A\gamma/^G\gamma$	α	HbF $\alpha_2{}^{A/G}\gamma_2$
Bone marrow	Adult	δ and β	α	HbA $\alpha_2\beta_2$ HbA$_2$ $\alpha_2\delta_2$

another. For humans, the ζ, ϵ, γ and α genes would all be active in embryonic erythroid cells derived from the yolk sac, but only the γ and α genes would be transcribed in foetal erythroid cells of liver origin; in adults, the δ, β and α genes would be expressed by differentiating red cells formed in the bone marrow. While true at a gross population level, this is by no means the whole story. A variety of experiments suggest that haemoglobin switching can also take place within single cells derived from the previous erythropoietic precursor population. This apparently occurs under the influence of circulating 'humoral' factors in the bloodstream. For example, human HbA and HbF can be detected within the same red cell at around the time of birth. Indeed, a small proportion of adult BFU-Es give rise to F cells containing some HbF as well as HbA, while most give A cells containing only HbA (see Nienhuis & Stamatoyan-nopoulos, 1978). Erythroid precursor cells can be transplanted into a recipient animal which has been lethally irradiated so as to destroy its own stem-cell population; under these conditions the erythropoietic sites are recolonised by transplanted cells, so that new erythrocytes are formed and the recipient survives. Zanjani *et al.* (1979) performed such a transplant from the livers of foetal sheep homozygous for the β^B variant, into irradiated adults homozygous for β^A (an alternative allele of the same β gene). Haemoglobin containing β^B chains was found to be expressed at increasing levels over a 45-day period following transplanta-tion, while production of HbF (containing foetal γ chains) remained very low. Since the stem cells transplanted were of foetal liver origin, they would normally give rise to erythrocytes expressing only HbF. But in the adult recipient environment, γ chain production is suppressed and β^B expression (donor marker) is activated in the progeny of the transplanted liver cells. This implies the existence of humoral factors circulating in the bloodstream which determine the pattern of globin expression, both switching off the old and switching on the new.

Similar evidence is available for the erythroid precursor cells of the embryonic yolk sac. Mouse cells of this type produce only primitive erythrocytes expressing embryonic haemoglobins when cultured in isolation. However, when cultured in transfilter combination with, for example, hepatic tissue from late mouse embryos (post-28-somite stage), these same yolk-sac precursors give rise to definitive erythrocytes expressing the haemoglobins characteristic of later development (Cudennec *et al.*, 1981). Chapman & Tobin (1979) used double fluorescence-labelling to show that both early and late haemoglobins are present simultaneously in single chick erythrocytes at around the time of changeover (day 6 of embryonic development). It is now thought that a single population of stem cells may migrate from one erythropoietic site to another during development.

Haemoglobin switching is regulated primarily at the transcriptional level. Groudine *et al.* (1981) demonstrated that embryonic chick erythroid nuclei synthesise no detectable adult globin transcripts; similarly no embryonic globin sequences could be detected in the nuclear RNA of adult chick erythroid cells – even under conditions which should block transcript processing and degradation.

The chromatin conformation of non-expressed globin genes has been probed extensively using DNase I. As mentioned in section 2.6, the foetal γ globin gene remains DNase I-sensitive in adult sheep erythroid cells, even though it is rarely if ever expressed there (but see note on A and F cells above). On the other hand, the non-expressed β genes are DNase I-resistant in foetal sheep erythroid cells (Young *et al.*, 1978). Thus the activation of β globin expression which occurs in sheep erythroid cells of foetal liver origin after transplantation into an adult recipient (see above), must presumably involve a change in chromatin conformation at the β gene site.

A converse situation applies in chickens. Here the non-expressed embryonic β globin genes are DNase I-resistant in adult erythroid cells. By contrast, both adult and embryonic types of β gene are DNase I-sensitive in embryonic erythroid cells; here the adult β gene is presumably in some kind of 'preactivation' state; since it is not yet expressed (Stalder *et al.*, 1980a). Weintraub *et al.* (1981) analysed the switching of α-type genes in chick, from U gene expression in primitive erythroid cells to α_D and α_A expression in definitive erythroid cells. During this transition, the U gene becomes DNase I-resistant and heavily methylated (see

§2.6), as well as losing a DNase I-hypersensitive site near its 5' end. In the definitive erythroid cells, a DNase I-sensitive undermethylated region extends from the 5' end of the α_D gene, through a 1.5 kbp spacer and the whole α_A gene, to end 1.5 kbp beyond the 3' end of the latter gene. Sharp boundaries mark both ends of this chromatin region. Mavilio *et al.* (1983) have recently shown in humans that undermethylated 5'-flanking sequences occur only at active gene sites among the β-type globin genes in embryonic (ϵ), foetal (γ) and adult (β) erythroid cells.

Groudine & Weintraub (1981) similarly examined the chromatin conformation of α-type and β-type globin genes in those regions of the 20–23 hour chick blastoderm which will later (at around 35 hours) initiate primitive erythropoiesis (i.e. the future blood-islands). They showed that even the primitive haemoglobin genes (such as U) are still DNase I-resistant and heavily methylated, showing no signs of transcriptional activity at this early stage. Thus a change in the chromatin conformation of these genes must occur between 23 and 35 hours of development in those cells destined to form primitive erythroid tissue.

5.4 Expression and control of globin genes

In erythroid cells a 15S precursor RNA (1500–1900 bases long) is transcribed from the entire split gene coding for β globin (see §3.9). This precursor is processed within the nucleus by removing both intron transcripts sequentially (small before large, each in two steps), so as to generate mature 9S β globin messenger RNA (Grosveld *et al.*, 1981b). All β-type globin genes possess one long and one short intron located at homologous positions within the gene, and their transcripts are presumably processed in a similar way. The α genes also have two introns at homologous sites, but both are short. As a result, the α primary transcripts are only 11S (about 880 bases long) and are again spliced sequentially to yield mature 9S α globin mRNA for export (see Curtis *et al.*, 1977). Little is yet known about processing for ζ gene transcripts, though Proudfoot *et al.* (1982) have shown that both introns in the human ζ gene are long.

Although translational controls have an important role in balancing the final levels of globins and other cell products (see below), the pattern of globin gene expression is controlled primarily at the level of transcription. This applies not only to globin gene switching (§5.3), but also to the onset of globin synthesis during differentiation. In chick embryos, globin proteins can be detected at low levels as soon as the corresponding mRNAs begin to accumulate during the proerythroblast stage (Hyer &

Chan, 1978). (Earlier, it had been thought that globin messenger appearance might precede the onset of translation, implying some form of mRNA storage or other post-transcriptional control.) The promoter sites recognised by pol II in the 5'-flanking regions of mammalian globin genes have already been discussed in section 3.4.

Somatic cell fusion has proved a useful technique for analysing the location and control of globin genes. In brief, when human and mouse cells are fused together in culture, the human chromosomes tend to be eliminated selectively, so that only one or a few of them are retained in particular sublines of hybrid cells. When such cells are analysed for the presence of the human α globin gene, positive hybridisation is found only in those sublines where human chromosome 16 is retained (Diesseroth *et al.*, 1977). Similarly, only hybrid cells retaining human chromosome 11 carry the human β globin gene (Diesseroth *et al.*, 1978). Thus the human α locus lies (together with ζ) on chromosome 16, and is unlinked to the β locus which lies (together with ϵ, γ and δ) on chromosome 11.

Willing *et al.* (1979) have shown that hybrids between human fibroblasts and mouse erythroleukaemia (MEL) cells can be induced chemically to express human globin mRNAs of α- or β-type, or even both together (depending on whether the hybrids retain human chromosomes 16 or 11 or both). However, human γ globin mRNA is never expressed in β-producing hybrids, even though the β and γ genes lie close together on chromosome 11 (see §5.5). The human fibroblast parents never express any of the globin genes, while the mouse parent cells (from a transformed erythroid line) are inducible for high-level expression of α and β but not γ globins. Thus a set of human genes, corresponding to those active in the mouse parent, can become activated in hybrid cells which retain the appropriate human chromosomes. This implies the existence of *trans*-acting positive regulatory factors specific to particular types of globin gene (γ is not activated), but able to work to some extent across species boundaries. However, the level of human globin-gene expression in these hybrids tends to be rather low and the corresponding proteins are not translated. Consistently higher levels of human α globin gene products (protein as well as mRNA) can be obtained in hybrids between MEL cells (as above) and human parent cells of erythroid rather than non-erythroid origin (Diesseroth *et al.*, 1980). These authors were able to select for hybrid cells retaining human chromosome 16 by means of a marker enzyme whose gene is closely linked to that for α globin in the human parent, but which is unavailable from the MEL parent; this particular gene product is essential for cell survival under certain culture conditions.

The results obtained with MEL–erythroid hybrids (high human α expression) as compared to MEL–non-erythroid hybrids (low human α expression) suggest additional *cis*-acting influences located within chromosome 16; these promote α globin gene activity only in erythroid cells. At least two controlling factors are implied by all this, one acting in *trans* and the other in *cis*, but the identities of both remain obscure.

More recently, human β-type globin genes have been introduced by cotransformation (§3.4) into TK$^-$ mouse erythroleukaemia cells. In stable TK$^+$ transformant lines, human β globin mRNA expression can be induced chemically along with mouse globin mRNAs (from the host cell genes). Human γ and ϵ globin mRNAs are *not* induced, and are only transcribed from a viral promoter adjacent to the TK gene. Thus correctly regulated expression of the introduced human β globin gene can be obtained in MEL cells; this applies both to fragments of human DNA containing the β globin gene plus 1.5 kbp of 5'-flanking sequence, and to larger fragments containing the entire β-type gene cluster (see fig. 58; Wright *et al.*, 1983).

Several feedback control mechanisms have been identified in red blood cells, some of which act as links between globin production and the haem biosynthetic pathway. For example, haem is found to inhibit the expression of δ-aminolaevulinate (DAL) synthetase, which is the key enzyme that initiates haem biosynthesis. This is a case of end-product repression; excess haem product will halt further haem synthesis until the free product molecules have been sequestered (as haemoglobin) through binding to new globin chains. Conversely, erythropoietin increases DAL synthetase activity and so stimulates haem production. This occurs concomitantly with the terminal divisions and differentiation promoted by erythropoietin among CFU-E cells (see §5.2 above). The haem precursor haemin specifically enhances α globin translation (Giglioni *et al.*, 1973), so bringing α and β production levels into the correct 1:1 balance. These are simple examples from an extensive network of metabolic controls which together ensure that differentiating red blood cells function efficiently and respond appropriately to changes in their environment.

5.5 Organisation of globin genes

As indicated in section 5.4, the β globin gene is located on human chromosome 11 while the α globin genes are on chromosome 16. Each is linked to other related globin genes; thus a 40 kb region of chromosome 11 includes one pseudogene together with the active ϵ, $^G\gamma$, $^A\gamma$, δ and β

genes, while a 27 kb region of chromosome 16 includes both α genes plus one ζ gene and two pseudogenes. A second β-type pseudogene, thought to lie on the 5' side of the ϵ gene (Fritsch *et al.*, 1980) now appears to be an artifact (Shen & Smithies, 1982). The mapping of these human globin genes has involved cloning long stretches of the chromosomal DNA containing them, followed by restriction and sequence analysis. The β globin gene cluster has been studied intensively in many laboratories, and the complete nucleotide sequences of all its active genes are now known (see Lawn *et al.*, Spritz *et al.*, Baralle *et al.*, Slightom *et al* and Efstratiadis *et al.*, 1980); an overall map of the β gene region is given in fig. 59A, based on Fritsch *et al.* (1980). Similar information is available for the human α gene cluster (Lauer *et al.*, 1980; Proudfoot *et al.*, 1982), and is summarised in fig. 59B.

The sequences and organisation of globin genes have been compared extensively between vertebrate species (see review by Jeffreys, 1982). The arrangement of the β-type genes in the order of their expression during development holds true for other mammals as well as for humans, although the number of genes and the intergenic distances vary considerably. In chickens, however, the two embryonic β globin genes (ϵ and ρ) are found flanking the two adult β globin genes in the order 5'–ρ–β_H–β–ϵ–3' (Dolan *et al.*, 1981). Thus the chick β-type globin genes are not arranged on the chromosome in the order of their expression during development. In *Xenopus* the α- and β-type genes are closely linked together, rather than segregated into separate clusters as in higher vertebrates (including chick).

Despite these differences in arrangement, the globin genes show several constant features. All vertebrate globin genes contain two introns

A β **globin gene cluster** (based on Fritsch *et al*, 1980)

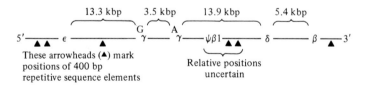

B α **globin gene cluster** (based on Lauer *et al*, 1980, Proudfoot *et al*, 1982 and Goodbourn *et al*, 1983)

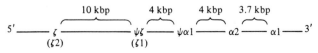

Fig. 59 Human globin gene organisation.

located at homologous positions within the coding sequence. In the case of the β-type genes, a short ~120 bp intron occurs close to the 5' end of the gene, while a longer intron (500–1000 bp in different mammals) occurs towards the 3' end. In the α globin genes, both introns are short, while in the human ζ gene both introns are long and partially composed of simple repetitive sequences (Proudfoot *et al.*, 1982). Despite these similarities in size and position, the actual sequences of the globin gene introns have diverged rapidly between species (van den Berg *et al.*, 1978). On the basis of their overall sequence relationships one can derive a hypothetical evolutionary tree showing how the present array of human globin genes could have arisen through successive duplication events from a single original globin gene (fig. 60), assuming a constant rate of mutational change and divergence. This is however an oversimplification, since mechanisms also exist for conserving the sequences of duplicated genes without significant divergence, as in the case of the two α gene copies. Thus the duplication times in fig. 60 are at best only provisional.

Several pseudogenes have been identified in human and other vertebrate globin gene clusters (see Little, 1982). They are presumably evolutionary relics, i.e. genes that were once functional but have now fallen into disuse (see also §1.9). Several globin pseudogenes have been sequenced; they are mostly found to retain the basic organisation of active globin genes (in terms of intron size, position etc.), but are mutated in various ways so as to preclude their expression. Examples of this include alterations in the TATA box sequence (§3.4), nonsense codons occurring early in the 'coding' region, and abnormal intron boundaries which would prevent transcript splicing. One β globin pseudogene (ψβ1) occurs in the human β-type gene cluster, and a rabbit pseudogene (ψβ2) related to the human δ gene has been partially sequenced (Lacy & Maniatis, 1980). In

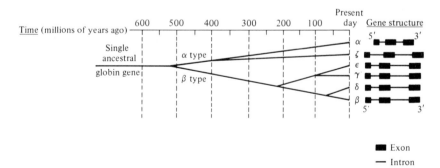

Fig. 60 Evolutionary tree for human globin genes (based on Alberts *et al.*, 1983, and Proudfoot *et al.*, 1982).

the human α globin gene cluster there is one α-type pseudogene ($\psi\alpha1$; Proudfoot & Maniatis, 1980) and one ζ-type pseudogene (here designated $\psi\zeta$; Proudfoot *et al.*, 1982); interestingly, the $\psi\zeta$ gene is almost identical in sequence to the active ζ gene, but codon 6 is mutated to a stop codon in $\psi\zeta$ (see Proudfoot *et al.*, 1982).

One exception to this general rule is the mouse $\psi\alpha3$ pseudogene, which represents an α-type gene sequence from which both introns are completely absent (Nishioka *et al.*, 1980; Vanin *et al.*, 1980). Since these intron sequences have been deleted precisely, there is an obvious analogy with transcript splicing. This has led to speculation that such 'processed genes' might originate from double-stranded cDNA copies of spliced RNAs, which have become reinserted into the genome in the germ line (Flavell, 1982; Little, 1982). Other examples of 'processed pseudogenes' include members of the tubulin, immunoglobulin and dihydrofolate reductase gene families. In the case of a dispersed human immunoglobulin pseudogene, the DNA sequence ends with a poly(dA) tract and includes spliced J and C regions (§4.2), both characteristic of processed immunoglobulin messenger RNAs (Hollis *et al.*, 1982). Nor are 'processed' genes necessarily inactive pseudogenes. In chick, one of the two active calmodulin genes contains introns whereas the other does not (Stein *et al.*, 1983); the latter, though possessing the characteristics of a processed gene, is nevertheless expressed in a tissue-specific manner in muscle cells.

The mouse genome is also unusual in that its two α globin pseudogenes ($\psi\alpha4$ with and $\psi\alpha3$ without introns) are dispersed onto two chromosomes different from that carrying the active cluster of one embryonic and two adult α globin genes (Leder *et al.*, 1981). If processed pseudogenes are indeed derived from spliced RNA intermediates, then dispersion to different chromosomal locations would be expected; but this does not explain the location of the intron-containing mouse $\psi\alpha4$ pseudogene on a different chromosome.

A variety of blood disorders affecting human globin genes and their expression have been analysed at the molecular level (see review by Weatherall & Clegg, 1979). In homozygous α-thalassaemia I, both α globin genes are deleted, hence only γ chains are available in the foetus; this is normally a lethal condition (*hydrops fetalis*). In $\delta\beta$-thalassaemia, both the δ and β genes are similarly deleted, again leading to severe anaemia. However, a more extensive deletion involving these same two genes is paradoxically much less severe in its effect; here, the absence of HbA/A$_2$ production can be compensated by a prolongation of HbF synthesis into adult life. Thus the $^G\gamma$ and $^A\gamma$ genes remain active in bone-marrow-derived erythroid cells, a condition known as hereditary persistence of foetal haemoglobin (HPFH). Possibly a site located between the $^A\gamma$ and δ genes (fig. 58A) is involved in the inhibition of $^{A/G}\gamma$ gene activity in adult erythroid cells; the disparity between $\delta\beta$ thalassaemia (severe effects) and HPFH (mild effects thanks to HbF compensation) might then be due to deletion of this site in the latter but not in the former case. However, at least one further site unlinked to the β-type globin genes is also involved in regulating γ gene activity in adults (see Gianni *et al.*, 1983).

In β thalassaemias, the adult erythrocytes contain little or no β globin protein. Usually, the levels of β mRNA are reduced or zero, though the β gene itself may remain largely intact, with no apparent deletions. In some of these cases, the defect has been traced to point mutations which alter the splicing pattern of the β globin primary transcripts (see Treisman *et al.*, 1982; Orkin *et al.*, 1982; Mount & Steitz, 1983). Promoter or nonsense mutations could also give the same end result.

The δ and β genes are fused together in Hb Lepore, due to deletion of the sequences that normally separate these genes (see fig. 58A). The result is a globin protein comprising part of a δ chain (lacking its C-terminus) joined to part of a β chain (lacking its N-terminus). Thus the 5' portion of the δ gene has become fused to the 3' portion of the β gene. In Hb Kenya, a much larger deletion results in fusion between the 5' portion of the $^A\gamma$ gene and the 3' portion of the β gene. These fusion mutants must have arisen originally by unequal crossing-over between sites within related globin genes (δ and β or $^A\gamma$ and β); for details see Weatherall & Clegg (1979).

If both globin gene clusters are taken as a whole, it is apparent that only some 8% of the available DNA sequences code directly for protein, while a similar percentage is represented by the introns (Jeffreys, 1982). The function of the rest (84%!) remains obscure, although the presence of interspersed repetitive elements has been noted (see fig. 58 and Fritsch *et al.*, 1980). It is also possible that the globin gene promoter sites may include sequences located at considerable distances from the genes whose expression they control (§3.4). Thus the global problems of DNA sequence organisation, function and significance discussed in chapter 1 (also §4.9) are here seen in microcosm, posing the same dilemmas and questions.

6

Egg production

Summary

In laying hens, the egg-white proteins are secreted by the magnum portion of the oviduct, whose growth and differentiation is dependent on female sex steroids (§6.1). The expression of the egg-white protein genes is also controlled by these hormones. Specifically, vast amounts of the corresponding mRNAs and proteins are rapidly induced by appropriate steroids in already differentiated oviduct tissue (secondary response), but this high-level production ceases during hormone withdrawal. However, the ovalbumin gene at least remains in a DNase I-sensitive chromatin conformation during withdrawal (§6.2). Differences in the lag-time between hormone administration and the onset of mRNA and protein accumulation are discussed in section 6.2. In particular, the conalbumin gene shows markedly different responses to oestrogen and to progesterone, suggesting complex hormonal controls.

Section 6.3 reviews the structure and expression of each of the major egg-white protein genes; namely those encoding lysozyme, ovomucoid, conalbumin, ovalbumin, and also two minor ovalbumin-related products from genes designated X and Y. Presumably DNA or chromatin sites adjacent to these genes are recognised by the hormone–receptor complex within the nucleus (see §4.8). Short consensus DNA sequences recur in the 5′-flanking regions of most or all of the egg-white protein genes, as discussed in the last part of section 6.3. Some of these short sequences probably represent *in vivo* binding sites for appropriate hormone–receptor complexes.

The yolk precursor proteins (vitellogenins) are synthesised in large quantities by the livers of female oviparous (egg-laying) vertebrates, again under hormonal control (§6.4). Injection of oestrogen into males (e.g. roosters or male *Xenopus*) results in the rapid induction of vitellogenin synthesis in the liver. This represents *de novo* hormonal activation of a gene never normally expressed in the male. In *Xenopus* there is a small family of four related vitellogenin genes (§6.4). The A1 and A2 members of this family are 95% identical in their exon sequences, but their introns (33 in each gene!) have diverged widely both in length and in sequence.

6.1 Introduction

That familiar foodstuff, the hen's egg, is a remarkable structure. Its two main constituents – the yolk and white – represent an enormous synthetic investment, in that gram quantities of their respective proteins are produced daily by a laying hen. For example, an average rate of egg production will involve the synthesis of 3×10^{19} molecules per day of the major egg-white protein ovalbumin (Palmiter, 1975). As mentioned in chapter 4, yolk platelets are assembled in the growing oocytes, but their component proteins are synthesised as longer precursors in the liver under hormonal control (see §6.4).

By contrast, the egg-white proteins are produced in the magnum portion of the hen oviduct, and are laid down as layers of albumen (thick, then thin) around the yolk during its passage down the oviduct. Shell membranes enclose each egg as it passes through the isthmus of the oviduct, and finally the shell (mainly calcium carbonate plus some brown porphyrin pigment) is secreted in the shell-gland prior to laying. The approximate times involved for each stage are as follows: (i) 3 hours in the magnum, (ii) $1\frac{1}{4}$ hours in the isthmus, and (iii) 20–21 hours in the shell gland, pigment being added only during the last 5 hours (Sturkie & Mueller, 1976).

The immature oviduct of a young female chicken is a narrow tube weighing only milligrams, composed mainly of a layer of columnar epithelial cells surrounded by connective tissue (fig. 61*A*). In a mature laying hen, however, the oviduct is a massive structure weighing 50–60 grams, of which at least half is contributed by the magnum. Although the ovaries and oviducts develop initially as paired structures in birds, those on the right side degenerate before sexual maturity, so that only the left ovary and oviduct are functional in laying hens.

Basically, the mature magnum is a thick-walled tube, whose luminal surface is intricately convoluted into ramifying *tubular glands*. Groups of tubular gland cells synthesise vast quantities of the major egg-white

Fig. 61*A*, *B*, *C* Differentiation of chick oviduct. Electron micrographs showing immature (7 week; part *A*) and mature (50 week; parts *B* & *C*) hen oviduct tissue, by courtesy of Dr I. R. Duce (Department of Zoology, University of Nottingham).

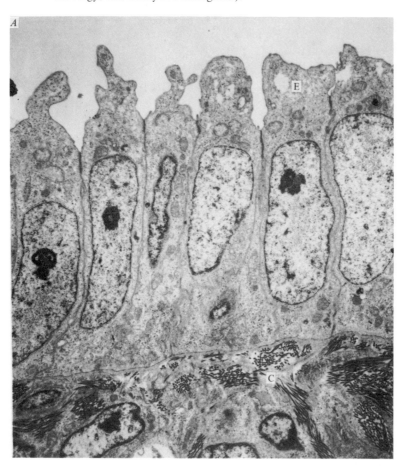

Fig. 61*A* *Immature oviduct*. Magnification ×2655. Note single layer of columnar epithelial (E) cells (relatively undifferentiated), with underlying connective (C) tissue (containing collagen fibre bundles).

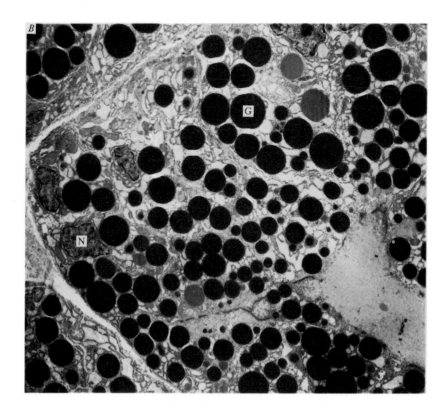

Fig. 61*B   Mature oviduct magnum – tubular gland cells*. Magnification
×1240. Note the electron-dense (dark) granules (G) of egg-white
proteins in the cytoplasm, and nucleus (N).

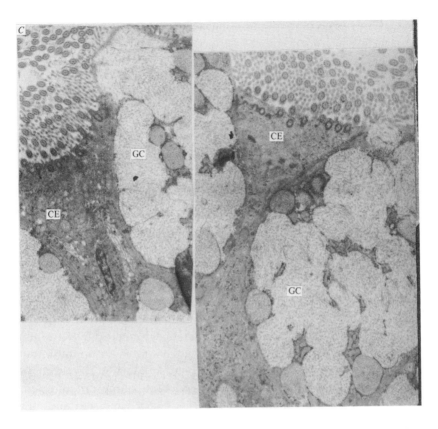

Fig. 61C *Mature oviduct – luminal lining layer*. Magnification ×1170.
Note the regular alternation of ciliated epithelial (CE) cells with goblet
(GC) cells packed full of mucopolysaccharides (speckled areas).

proteins (principally ovalbumin, but also conalbumin, ovomucoid and lysozyme), which are assembled into granules (fig. 61*B*) for secretion into the lumen. Ovalbumin is the major storage material in the egg white, while conalbumin acts as an iron-carrier and lysozyme as a bactericidal agent. Although tubular gland cells predominate numerically in the magnum (by about 9 to 1), two other cell types are found in an alternating arrangement in the layer lining the lumen. These are: (i) the goblet cells, which secrete avidin (see below) and also mucopolysaccharides to lubricate the egg's passage; and (ii) ciliated epithelial cells, which assist in moving the secreted material (fig. 61*C*). The outer wall of the oviduct is composed of connective tissue and smooth muscle, whose contractions are mainly responsible for moving the large egg down the oviduct.

It will be apparent from the contrasts in fig. 61 (*A* versus *B* and *C*) that extensive growth and differentiation are required in order to convert a simple immature oviduct into the complex adult organ. Most of this occurs during sexual maturation at around 100 days post-hatching. The whole process is under hormonal control, as is the subsequent synthesis of egg-white proteins. The major hormones involved are the female sex steroids, i.e. oestrogens (mainly 17β-oestradiol) and progestins (mainly progesterone). Both of these steroid classes are produced in the ovary, and are themselves subject to hormonal regulation from the pituitary. The interactions between oestrogen and progesterone are complex, partly reinforcing, partly modulating, and partly antagonising each other's effects in the oviduct (see table 10 later). For instance, oestrogen is prerequisite for the differentiation of goblet cells in the luminal lining of the magnum, but only when further stimulated by progesterone can these cells secrete avidin (a minor egg-white protein that binds biotin). Another important distinction is that oestrogen indirectly promotes mitosis among oviduct cells, whereas progesterone does not. The egg-white protein genes in differentiated oviduct cells can also respond to other steroids, namely androgens and glucocorticoids.

When an immature hen is first exposed to oestrogen (either naturally or after precocious injection of hormone), a complex *primary response* is elicited in the oviduct. This involves massive cell proliferation and tissue growth, accompanied by the differentiation of several distinct cell types (at least three in the magnum alone; fig. 61), and regionalisation of the oviduct into the infundibulum, magnum, isthmus and shell gland. Although egg-white protein synthesis is induced during this primary response, most workers have preferred to study the simpler secondary response. If oestrogen is withdrawn after the primary response, production of the egg-white proteins ceases, but the differentiated oviduct structure does not regress. When hormone is readministered, egg-white protein synthesis is rapidly reinduced without the necessity for preliminary differentiation. This *secondary response* is both faster and more extensive (i.e. higher production of egg-white proteins) than the primary response. Successive cycles of hormone administration and withdrawal give a reproducible pattern of induction and repression of egg-white protein expression. Eventually, however, the oviduct becomes post-reproductive and can no longer respond to further hormonal stimulation. In the account below (§6.2), we shall concentrate specifically on the expression of egg-white proteins in the tubular gland cells of the oviduct magnum during secondary hormonal responses.

6.2 Control of egg-white protein synthesis

During secondary oestrogen stimulation, ovalbumin production accounts for around 60% of total protein synthesis in oviduct tubular gland cells, while the production of conalbumin, ovomucoid and lysozyme reaches 10%, 8% and 2–3% respectively. In the resting oviduct during hormone withdrawal, these proteins are no longer synthesised, and their corresponding mRNAs are virtually undetectable (less than 0.0002% of total cellular RNA). 12 hours after hormone administration, ovalbumin mRNA represents 0.1% of the total cellular RNA, while ovomucoid and lysozyme mRNAs represent 0.0065% and 0.005% respectively (Schutz *et al.*, 1977). At this stage the concentrations of all three messengers are still rising. By contrast, the levels of conalbumin mRNA plateau after 6 hours at around 0.02% of the total cellular RNA (approximate value based on Palmiter *et al.*, 1977). Since the vast bulk of cellular RNA is ribosomal, these values represent massive new messenger synthesis. In the case of ovalbumin mRNA, for instance, the resting level of 10–30 molecules per tubular gland cell rises to at least 10 000 per cell by 12 hours after hormone administration, and to 70 000 per cell during maximal stimulation (table 11). The major egg-white protein mRNAs increase in concentration by more than a thousand fold in response to oestrogen.

Conalbumin mRNA is induced by oestrogen much more rapidly than the other egg-white protein messengers (see table 9, fig. 62 and discussion below). The time of first appearance for the corresponding proteins suggests that translation begins as soon as the mRNAs reach the cytoplasm; i.e. the kinetics of mRNA and polypeptide appearance are practically indistinguishable. Thus the temporal expression of all four egg-white proteins is mainly controlled at the level of mRNA production; to some extent this is also true in quantitative terms (compare columns 2 and 3 in

Table 9. *Control of egg-white protein expression (after Schutz* et al., *1977 and* Palmiter *et al., 1977).*

Protein	% of total cell protein synthesis in laying hen oviduct magnum	mRNA, as % of total cell RNA 12 hours after oestrogen administration	mRNA, as % of total cell RNA in resting oviduct	Time-lag between oestrogen injection and mRNA appearance (hours)
Ovalbumin	50–65%	0.1%, rising	0.0002%	2–3
Conalbumin	10%	approx. 0.02%, plateau	approx. 0.0002%	0.5
Ovomucoid	8%	0.0065%, rising	0.0001%	2–3
Lysozyme	2–3%	0.005%, rising	0.0001%	2–3

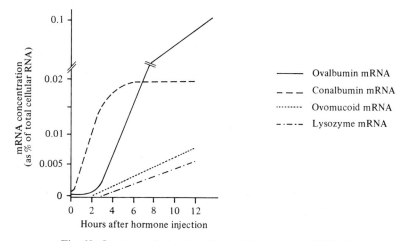

Fig. 62 Oestrogen induction of egg white protein mRNAs (after Schutz *et al.*, 1977, and Palmiter *et al.*, 1977).

table 9). Moreover, nuclear precursors to these mRNAs are barely detectable during hormone withdrawal, suggesting that all four genes are regulated transcriptionally. Another important effect of the hormone is to stabilise the egg-white protein messenger RNAs specifically. Many other mRNA sequences are expressed at low levels in hormone-stimulated oviduct cells (see §3.10 and fig. 40), and these continue to be expressed during hormone withdrawal. Thus secondary hormone administration elicits high-level expression of a limited set of egg-white protein genes, this effect being reinforced by a stabilisation of their messengers. These data are summarised in table 9 and fig. 62.

Table 10. *Induction of conalbumin and ovalbumin mRNAs by oestrogen and progesterone (after Palmiter et al., 1977).*

Hormone	Time between hormone injection and first appearance of mRNA (hours)		Initial rate of mRNA accumulation (molecules/min./cell)	
	Conalbumin mRNA	Ovalbumin mRNA	Conalbumin mRNA	Ovalbumin mRNA
17β-oestradiol	0.5	2–3	20	20
Progesterone	2–3	2–3	8	20
17β-oestradiol + progesterone (administered simultaneously)	2–3	2–3	12	32

The 2–3 hour time-lag between oestrogen administration and messenger appearance for the ovalbumin, ovomucoid and lysozyme mRNAs, contrasts sharply with the rapid induction of conalbumin mRNA which appears within 30 minutes of oestrogen injection. It will be recalled from section §4.8 that steroid hormones first bind to specific receptor proteins found only in the target cells, and that the activated hormone–receptor complexes then alter the pattern of gene expression. Several types of steroid receptor protein are found in oviduct tubular gland cells; these include not only oestrogen and progestin receptors, but also glucocorticoid and androgen receptors. All of these hormones can induce eggwhite protein synthesis as a secondary response in differentiated oviduct cells, though only oestrogens can induce the primary response. However, the pattern of gene expression is different for each of these hormones. Following administration of radioactively labelled steroid, maximal levels of the oestrogen–receptor complex are found inside the nuclei of oviduct cells within 30 minutes of hormone injection, implying that conalbumin gene transcription must be switched on almost immediately by this complex. The delayed expression of the other egg-white protein genes may possibly reflect some requirement for the hormone–receptor complex to translocate along the nuclear chromatin (see Palmiter *et al.*, 1976); alternatively, other control processes may mediate in the activation of these genes. If progesterone is used in place of oestrogen, a distinctly different pattern of conalbumin induction is observed (see table 10); in particular, there is a pronounced 2–3 hour lag before its mRNA first appears. Indeed, if progesterone is administered during the course of

oestrogen stimulation, then conalbumin mRNA accumulation is temporarily halted while ovalbumin mRNA synthesis continues (see Palmiter *et al.*, 1977).

These complexities apparently reflect direct interactions between the hormones and their target cells, since identical results are obtained by treating oviduct explants (from oestrogen-stimulated birds) with one or both steroids *in vitro* (McKnight, 1978).

In view of the massive amounts of ovalbumin produced by the tubular gland cells in a laying hen, it is worth noting that the major ovalbumin gene is single-copy and is *not* amplified before or during hormone stimulation. Palmiter (1975) has reviewed the transcriptional and translational parameters which permit the production of 3×10^{19} ovalbumin molecules per day in a laying hen's oviduct. There is some uncertainty in the figures quoted in part *C* of table 11, due to a very rough estimate of the number of tubular gland cells (probably $2\text{--}3 \times 10^{10}$) in each oviduct magnum.

Table 11. *Parameters of ovalbumin production in tubular gland cells (after Palmiter, 1975).*

A Transcription rate		B Translation rate	
Parameter	Value	Parameter	Value
Initial rate of mRNA$_{ov}$ accumulation (after oestrogen stimulation)	20 mols/min./cell	Polysome size	12 ribosomes/mRNA$_{ov}$
Steady-state rate of mRNA$_{ov}$ synthesis in laying hens	34 mols/min./cell	Ribosome transit time (i.e. time taken to complete one ovalbumin protein chain)	1.3 min.
Half-life of mRNA$_{ov}$	24 h	Rate of protein elongation	300 amino acids/min.
Mean life-time of each mRNA$_{ov}$ molecule	35 h	Rate of ribosome initiation on each mRNA$_{ov}$ molecule	9.2 ribosomes/min.

C Steady state concentration of mRNA$_{ov}$ (a) in oestrogen stimulated oviduct, 70 000 mols/cell
(b) in laying hen oviduct, 100 000 mols/cell
Rate of ovalbumin synthesis: 6.4×10^5 mols/min./cell
Assuming $2\text{--}3 \times 10^{10}$ tubular gland cells per oviduct, this gives 3×10^{19} mols/day/hen

Following hormone withdrawal, transcription of the egg-white protein genes virtually ceases, and presynthesised cytoplasmic mRNAs are degraded. Indeed, ovalbumin mRNA has a markedly shorter half-life during acute hormone withdrawal than during stimulation, suggesting that this messenger is stabilised in the presence of hormone (Palmiter & Carey, 1974). Stocks of preexisting egg-white proteins also disappear from the lumen of the tubular gland in the absence of hormone. However, as mentioned in section 2.6, the chromatin of the ovalbumin gene remains DNase I-sensitive during hormone withdrawal (Palmiter *et al.*, 1977). Presumably this facilitates the rapid reinitiation of transcription from the ovalbumin gene when a further hormonal stimulus is given. The 5′-flanking sequences of the ovalbumin gene in oviduct chromatin contain an extensive region of several hundred bp which is highly sensitive to DNase I, rather than the usual pattern of shorter DNase I-hypersensitive sites (Gross-Bellard *et al.*, quoted in Igo-Kemenes *et al.*, 1982).

6.3 Egg-white protein genes and their organisation

The egg-white protein genes in chick are each subdivided by several introns, ranging in number from 3 (lysozyme gene) to 16 (conalbumin gene), and each gene is 4–6 fold longer than its mRNA. The structures of these genes are summarised briefly below, followed by a more tentative discussion of the 5′-flanking sequences which may play some role in their regulation by steroid hormones.

(i) *Lysozyme gene.* This gene occupies 3.9 kbp of chromosomal DNA and includes 3 intron sequences (Lindenmaier *et al.*, 1979). All of these occur in the protein-coding portion of the gene, and the four exons correspond approximately to distinct domains of the protein structure (Jung *et al.*, 1980).

(ii) *Ovomucoid gene.* Seven introns have been identified within a genomic DNA fragment representing the entire chicken ovomucoid gene (Catterall *et al.*, 1979; Stein *et al.*, 1980). Hen oviduct nuclear RNA contains a series of ovomucoid mRNA precursors ranging in size from 1.5- to 5-fold longer than the final messenger (Nordstrom *et al.*, 1979). The largest of these species (5.5 kb) is probably a primary transcript of the entire gene (cf. §3.9); all of the precursors carry a poly(A) tag, and the removal of intron transcripts follows a preferred order during processing (Tsai *et al.*, 1980). Once again there is evidence that each exon more or less corresponds to a functional domain of the ovomucoid protein; moreover, the present gene appears to have evolved via duplication of a DNA sequence that was already split (see Stein *et al.*, 1980).

(iii) *Conalbumin gene*. This gene is considerably larger and more complex, being subdivided by 16 introns into 17 exons (some as short as 60 bp; Cochet *et al.*, 1979). Gene duplication is again implicated in the evolution of this structure. Repetitive sequences occur both in the 5'-flanking regions and within one of the introns of the conalbumin gene. Oviduct conalbumin is in fact the same protein as liver transferrin (which acts as an iron-carrier in the blood). These two proteins have identical amino-acid sequences, but differ in their carbohydrate moieties attached post-translationally. The regulation of their expression, presumably from one and the same gene, is quite different in the two tissues involved. Thus liver production of transferrin is regulated in part by iron levels in the blood, but is much less responsive to the steroid hormones which induce conalbumin synthesis in the oviduct. Nor is this due to a lack of steroid receptors in liver cells, since vitellogenin production in the liver is strongly induced by oestrogen (see §6.4 below). This implies complex control mechanisms which for the present remain to be clarified. (See also discussion in section 6.2 above on the differential response of conalbumin expression to progesterone as compared to oestrogen.)

(iv) *Ovalbumin gene*. The ovalbumin gene in chickens is split by seven introns (300–1400 bp long) into eight exon segments (Dugaiczyk *et al.*, 1979). The ovalbumin messenger RNA contains an unusually long 3' untranslated region, and this part of the mRNA-coding sequence is uninterrupted in the genomic DNA. Six of the introns occur within the protein-coding region of the gene, while the seventh (which is long) interrupts the DNA coding for the 5' untranslated region of the messenger (fig. 63; see Breathnach *et al.*, 1978). This delineates a short 5' exon encoding a signal sequence in the mRNA.

The entire gene of 7.7 kbp is transcribed into a long precursor RNA from which the seven intron transcripts are removed in a preferred order within the nucleus (Tsai *et al.*, 1980). The final messenger is 1859 bases long [excluding the cap and poly(A)] and has been completely sequenced

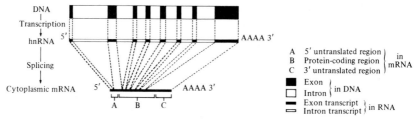

Fig. 63 Structure and expression of the ovalbumin gene (not to scale)
From Breathnach *et al.* (1978) and Dugaiczyk *et al.* (1979).

(McReynolds *et al.*, 1978). The presence of consensus sequences at exon–intron and intron–exon junctions within the ovalbumin gene has already been discussed (see Catterall *et al.*, 1978, and Breathnach *et al.*, 1978).

Several other points concerning this gene have been mentioned earlier. The chromatin of the ovalbumin gene is preferentially digested by DNase I in hen oviduct nuclei (Garel & Axel, 1976), but not in nuclei from other (e.g. erythroid) tissue sources. This sensitivity is maintained in oviduct cells even during hormone withdrawal (Palmiter *et al.*, 1977; see §6.2). Finally, the entire chick ovalbumin gene has been introduced into mouse TK$^-$ cells by cotransformation with the Herpes virus TK gene (see §3.4). Some TK$^+$ transformants are found to express high levels ($1\text{–}100 \times 10^3$ molecules/cell) of immunologically recognisable ovalbumin protein (Lai *et al.*, 1980), implying correct transcription, processing and translation of the foreign gene.

(v) *Ovalbumin-related X and Y genes*. Royal *et al.* (1979) have mapped two large cloned segments of chicken DNA which overlap for about 1000 bp within the ovalbumin gene. One of these, 17 kbp long, extends in a 3′ direction beyond the ovalbumin gene; the other, 30 kbp long, extends in a 5′ direction from the ovalbumin gene, so covering some 46 kbp of DNA in all. Within the DNA on the 5′ side of the ovalbumin gene, Royal *et al.* (1979) identified two further genes related both in sequence and in intron/exon pattern to that encoding ovalbumin. These two relatives of the ovalbumin gene are designated X and Y, with Y located between the X and ovalbumin genes; stretches of non-expressed DNA sequences separate these three genes. All three lie in the same orientation, forming a small family of linked genes rather reminiscent of the globin gene-clusters (chapter 5). Both X and Y appear to be functional genes (not pseudogenes), X being transcribed to give a 2400-base messenger after processing, while Y gives a 2000-base mRNA. Neither of these mRNAs is abundant, yet both are synthesised under the same conditions of hormone stimulation as the ovalbumin messenger (see also LeMeur *et al.*, 1981). The precise function of the X and Y genes remains unclear, but probably they encode minor egg-white proteins.

(vi) *Coordinate hormonal control of the egg-white protein genes*. The above account indicates that at least six genes (X, Y, ovalbumin, ovomucoid, lysozyme and conalbumin) respond to hormonal induction in chick oviduct cells. A variety of different steroids can elicit similar responses, including androgens and glucocorticoids as well as progestins and oestrogens. Induction is coordinate in that the same set of genes is affected by all these hormones, although the resulting patterns of gene expression are somewhat different in each case (§6.2 and table 10). This suggests additional controls governing the response of each individual egg-white protein gene.

One may ask whether the various types of hormone–receptor complex recognise specific DNA sequences in the 5'-flanking regions of these genes, or whether their binding is mediated e.g. by chromatin proteins. Detailed sequence analysis of the 5'-flanking regions, together with studies of hormone–receptor binding to fragments of cloned DNA, suggest that short sequence elements fulfilling such a function are shared by many if not all of the egg-white protein genes.

Mulvihill *et al.* (1982) reported that a crude preparation of the progesterone receptor binds preferentially to cloned DNAs at sequences located 250–300 bp upstream from the 5' ends of five of these genes (the lysozyme gene was not included). Partial deletion mutants lacking these 5'-flanking sequences showed no preferential binding. The following consensus sequence was identified within the progesterone receptor-binding regions of all five genes:

$$5'\text{-ATC}^{CC}_{TT}\text{ATT}^{A}_{T}\text{TCTG}^{G}_{C}\text{TTGTA-}3'$$

However, Compton *et al.* (1983) have described AT-rich sequences lying closer to the 5' ends of the ovalbumin and Y genes, which preferentially bind a highly purified preparation of the progesterone receptor. It seems likely that multiple DNA sequences with binding affinity for hormone–receptor complexes may occur upstream from and even within these genes (see Parker, 1983). However, not all of these binding sites necessarily function as such *in vivo*.

One technique for distinguishing essential from inessential binding sites involves a modified type of cotransformation assay. Various lengths of cloned DNA representing the 5'-flanking sequences from a hormone-responsive gene are covalently linked to defined coding segments of foreign DNA (e.g. a mammalian globin gene), and these 'chimaeric genes' are then introduced into cultured oviduct cells. Hormone-responsiveness can then be gauged by comparing the levels of chimaeric gene expression both before and after induction with appropriate hormones. Using this approach, Renkawitz *et al.* (1982, 1984; see also Parker, 1983) have shown that sequences located between 164 and 208 bp upstream from the start of the lysozyme gene are necessary for progesterone induction of expression. Similarly, sequences located between 95 and 220 bp upstream from the ovalbumin gene are required for progesterone-induced expression (Dean *et al.*, 1983).

There is evidence from other hormone-responsive systems that the DNA sites recognised by hormone–receptor complexes may lie within transcriptional enhancer sequences. An example of this occurs in the long terminal repeats (LTRs) of mouse mammary tumour virus, whose expression is regulated by glucocorticoid hormones (see Parker, 1983, and Scheidereit *et al.*, 1983). Whether this is also true for hormonal induction of normal cellular genes remains to be seen. Short DNA sequences recur in the 5′-flanking regions of all the chicken egg-white protein genes (as discussed above), and this is also true for a number of rat and human genes responsive to glucocorticoids (see Cochet *et al.*, 1982). Some of these shared repetitive sequences are probably involved in hormone–receptor binding, but others could conceivably function as 'receptor' or 'identifier' elements involved in some other way in the integrated control of hormone-responsive gene-batteries. This speculative question has been touched on in section 4.9 (see also Davidson *et al.*, 1983).

6.4 Vitellogenin genes and their regulation

Within the oocyte, yolk platelets are assembled from fragments of the large precursor proteins known as vitellogenins (MW 170 000), as previously discussed in section 4.4. Vitellogenins are actually synthesised in the female liver in oviparous vertebrates, and are transported to the ovary as dimers in the bloodstream. Like the egg-white proteins in oviduct, vitellogenin expression in liver is under hormonal control by female sex steroids.

6.4.1 Chicken vitellogenin

Oestrogen levels are normally very low in roosters, and vitellogenin mRNA is almost undetectable in the liver (approx. 0.5 molecules per cell). However, even male liver cells contain oestrogen receptors; (similarly, the egg-white protein genes can respond to androgens in female oviduct). Thus if 17β-oestradiol is injected into roosters, vitellogenin synthesis is induced rapidly in the liver. Note that this is a *primary* response to the hormone, and is much easier to study than the primary response in oviduct since the liver cells are already differentiated. Vitellogenin mRNA levels reach a maximum of around 6000 molecules per cell within 3 days of hormone injection (Burns *et al.*, 1978). Thereafter the levels of this messenger decline exponentially (half-life of 29 hours), falling to less than 10 molecules per cell by the 17th day after injection.

Another yolk protein known as VLDL II (very-low-density lipoprotein) is also synthesised by hen liver cells *in vivo*, and shows a similar pattern of oestrogen response in rooster liver cells. Increased transcription of both VLDL II and vitellogenin genes is detectable within 30 min. of a single hormone injection (Wiskocil *et al.*, 1980). After reaching maximum levels of several thousand molecules per cell, new transcription ceases, and preexisting mRNAs of both types disappear over a period of days; this may involve destabilisation of the mRNAs in the absence of hormone (cf. Palmiter & Carey, 1974, discussed in §6.2 above). According to Wiskocil *et al.* (1980), the levels of mRNA encoding a normal male liver protein (serum albumin) are unaffected by the oestrogen treatment. However, other workers had earlier reported that serum albumin mRNA levels decline significantly during oestrogen stimulation of rooster liver (Williams *et al.*, 1978).

As mentioned in section 2.6, primary oestrogen treatment causes three new DNase I-hypersensitive sites and a single demethylated site to appear in the 5′-flanking region of the major chicken vitellogenin gene (VTG II; see Burch & Weintraub, 1983). The latter and two of the former remain present during hormone withdrawal, so that secondary oestrogen treatment induces only a single DNase I-hypersensitive site in this region. This may explain why liver cells respond more rapidly to secondary stimulation by oestrogen, and give higher levels of vitellogenin expression than during the primary response (cf. egg-white protein production in oviduct cells; §6.1 above).

6.4.2 Xenopus *vitellogenins*

A similar pattern of primary and secondary responses involving vitellogenin induction by oestrogen has been described for *Xenopus* male liver cells in culture (Searle & Tata, 1981). This system facilitates the analysis of early events in the induction process. *Xenopus* vitellogenin mRNAs are also much more stable in the presence of oestrogen (half-life 3 weeks) than in its absence (half-life 16 hours; Brock & Shapiro, 1983). Thus oestrogen exerts a dual control over vitellogenin levels, acting both on transcription and on mRNA stability.

The existence of a small family of four related vitellogenin genes in *Xenopus* was shown using cloned cDNA sequences derived from the 6.3 kb vitellogenin messengers (Wahli & Dawid, 1979). The mRNA-coding sequences (exons only) of the two B-type genes have diverged in about 20% of their nucleotide residues from those of the two A-type genes. By contrast, the two A-type genes differ from each other in only 5% of their exon sequences, and the same is true for the two B-type coding regions. This implies an evolutionarily ancient duplication of the ancestral vitellogenin gene to give the A and B classes, each of which has subsequently undergone a further duplication to yield the A1 and A2 plus B1 and B2 genes. While one major class of vitellogenin proteins is processed into phosvitin and lipovitellin I + II fragments (see §4.4), a different class of vitellogenin polypeptides gives lipovitellin I + II and phosvette I + II fragments (these last being small yolk phosphoproteins of MWs 14000 and 19000 respectively). Thus the evolutionary divergence of the vitellogenin A and B genes is reflected in a functional diversification among their products (see Wahli *et al.*, 1981; Wiley & Wallace, 1981).

Wahli *et al.* (1980) have compared the genomic organisation of the two class A vitellogenin genes. Both are highly complex genes split by 33 introns at homologous positions (see §3.8 and fig. 34). As mentioned above, the exon sequences (mean length 175 bp) are very similar in both genes, differing in only 5% of their base pairs. Perhaps surprisingly, the two genes are quite different in overall length; 21 kbp for A1, but only 16 kbp for A2. This difference is due entirely to the greater lengths of the introns in the A1 gene (mean intron length 450 bp) as compared to the A2 gene (mean intron length 310 bp). Whereas the exons have evolved slowly through point mutations, the introns have evolved much more rapidly and radically – with examples of deletions, insertions and duplications as well as point mutations (Wahli *et al.*, 1980).

At least six middle-repetitive sequence-families are represented among the introns of the A1 gene (Ryffel *et al.*, 1981). During oestrogen stimulation, they are of course transcribed as part of the vitellogenin nuclear precursor RNA. They can therefore be classed as nucleus-confined repetitive transcripts. Interestingly, these same six repetitive sequences are also expressed in the nuclear RNA of male liver cells never exposed to oestrogen, i.e. where the vitellogenin genes are silent (Ryffel *et al.*, 1981). Thus members of all six repetitive families must also occur within other transcription units (genes) that are *not* regulated by oestrogen. Until the identity and control of these 'other genes' have been elucidated in detail, any relationship with the Britten–Davidson model (§4.9) remains in the realms of speculation. Recently, three of these repetitive sequences have been identified within introns of the larger albumin gene in *Xenopus*. This gene is expressed in liver cells at all times, i.e. during hormone withdrawal as well as during stimulation. However, the shared repetitive sequence elements in both albumin and vitellogenin genes are found at scattered locations, which do not suggest any obvious regulatory role (Ryffel *et al.*, 1983). If further genes expressed only in liver are found to share one or more of the same repetitive elements, and if such sequences are absent from the nuclear transcripts in other cell types, then this system might provide another example of tissue-specific 'identifier' sequences (see end of §4.9).

7

Insect development

Summary

In holometabolous insects such as *Drosophila*, a complete reconstruction of the body is initiated by the hormone ecdysone during *metamorphosis* from the larval to the adult (imago) stage. The polytene larval tissues are broken down and most of the adult develops from reservoirs of diploid imaginal cells which have remained undifferentiated during larval life (§7.1). The patterns of transcriptional activity in larval salivary-gland cells can be visualised even under the light microscope, since decondensed regions termed *puffs* occur at particularly active sites along the length of the giant polytene chromosomes in this tissue (§7.2). Some puffs are active during larval life (e.g. the large puff designated BR2 in *Chironomus*; §7.2), but these mostly regress in response to ecdysone. This hormone also induces a set of new puffs, in two stages (§7.2), as a preparation for metamorphosis.

Ecdysone initiates the evagination and final differentiation of the *imaginal discs*. These are discrete groups of imaginal cells destined to form particular regions of the adult exoskeleton (§7.3). Determination occurs stepwise in *groups* of cells occupying precise regions within each imaginal disc; each such group will form a defined subsidiary element of the adult structure during metamorphosis. A change in the determined state (termed *transdetermination*) sometimes occurs during *in vivo* culture of imaginal discs under conditions where growth continues but no differentiation occurs (i.e. in the absence of ecdysone; §7.4). A transdetermined group of cells will give rise to an adult structural element inappropriate to the original disc-type, when allowed to metamorphose in the presence of ecdysone (e.g. wing tissue produced by cells of leg-disc origin).

The segment pattern characteristic of insects arises from discrete groups of blastoderm cells; each such group acquires a segmental specification or '*address*' during early embryogenesis (§7.5). The larval segments correspond in number and type to the adult segment pattern, although there is no simple one-to-one relationship between segments and imaginal discs. During larval life, each imaginal disc becomes subdivided step by step into a series of *compartments* (e.g. anterior/posterior, dorsal/ventral), which develop autonomously in that cells do not cross the compartment borders once established (§7.5). Sometimes compartment boundaries correspond to no obvious feature in either the disc or the adult structure derived from it, e.g. the anterior/posterior border running across the wing disc and wing. Pairs of compartments seem to be established by a simple binary determination decision taken by a group of cells together (forming a *polyclone*). During larval life, each imaginal cell thus acquires an increasingly detailed address which specifies its future development.

The 'selector gene' model (Garcia-Bellido, 1975) proposes that these decisions involve changes in the pattern of activity among a small number of regulatory (selector) genes. A binary choice between alternative pathways, for instance, might reflect an 'on' or 'off' state for one particular selector gene (see §§7.4 and 7.6). These genes would act together in a combinatorial fashion. Such genes are apparently defined by *homoeotic* mutations, which cause remarkable disturbances in the larval and adult body plan. Generally, structures characteristic of one segment or compartment are transformed in homoeotic mutants into structures characteristic of another segment or compartment (cf. transdetermination). The phenotypes and genotypes of several homoeotic mutants (e.g. those involving the *engrailed* locus and bithorax complex) are discussed in section 7.6. The clear implication is that segment-specific and compartment-specific patterns of activity among the wild-type homoeotic genes underlie the normal distinctions between segments and between compartments in the insect body. Several of the homoeotic genes in *Drosophila* share a common conserved sequence (the 'homoeo box'); related sequences are present even in vertebrate genomes, suggesting that common principles may underlie the genetic determination of body plan in a wide variety of animals.

7.1 Introduction

Drosophila remains the best-characterised of all animals in genetic terms
(§4.1), with a formidable array of mutations (maintained in laboratory
stocks) that affect various aspects of its development. Other advantages
include the short life cycle, the small size of the *Drosophila* genome
(§1.2), and the characteristic banding patterns of larval polytene chromosomes (see below).

The initial cleavage of the zygote nucleus to give a syncitial and then a
cellular *blastoderm* has been outlined in section 4.5 (fig. 49). Complex
foldings of the blastoderm (gastrulation) ensue to produce a multilayered
germ band, and the characteristic pattern of body segments becomes
apparent within this band by about 10 hours of development. The
segment pattern is a fundamental feature of insect organisation (7.5);
after hatching it persists throughout larval development, and is also
apparent in the adult (imago) stage following metamorphosis.

During larval life the majority of nuclei become *polytene* through
repeated DNA replication without separation of the daughter chromatids
(see also §4.2). Consequently numerous DNA duplexes (chromatids) lie
side by side in each polytene chromosome. In some larval tissues (particularly salivary gland), the high degree of polytenisation results in giant
chromosomes with characteristic banding patterns clearly visible under
the light microscope. The transverse *bands* represent a perfect alignment
of condensed 'chromomeres' along the length of each chromatid. Bands
contain the majority of chromosomal DNA, whereas the less-condensed
interbands contain only about 5%. There is a general correlation between
the number of cytological bands and the number of genetic loci identified
by classical linkage studies within any given chromosomal region (see e.g.
Judd & Young, 1973; Hochmann, 1973). At the chromatid level, however, each chromomere (representing one band) contains far more DNA
in linear terms than would be required for a single gene. Moreover, some
bands are now known to contain small clusters of genes with closely
related functions; examples include the 66D group of chorion protein
genes (§4.2) and the 68C locus described below (§7.2). It is thus an
oversimplification to equate one band with one gene site. The transcription (puffing) of genes within the bands will be described in section 7.2
below.

One consequence of this polytenisation process is that larval cells
cannot divide, but merely grow larger as development continues. The
larval stage is divided into several *instars* (three in *Drosophila*), separated
by *moults* when the exoskeleton is shed and regrown to accommodate

extra growth. Each moult is triggered by the steroid hormone ecdysone (§7.2), whose action is modulated during larval life by the *juvenile hormone*. The levels of both hormones are low during the intermoult periods, but rise and then fall cyclically during moulting.

At the end of their larval life, holometabolous insects (such as the Diptera) undergo *metamorphosis*, which results in a radical change of body form. This process is triggered by rising levels of ecdysone, but only in the absence of a challenge from juvenile hormone. The final larval instar secretes a special *puparium*, within which the transformation from larva to adult (imago) takes place during the *pupal* stage. The polytene larval tissues are not simply remodelled during metamorphosis; rather they are discarded wholesale and broken down. The new structures of the adult develop instead from groups of diploid *imaginal cells* which have been present throughout larval life, though playing no active part in larval development. These cells are set aside in small groups during early development, and divide repeatedly without overt differentiation during larval life. The imaginal cells are present in the larva both as *imaginal discs* (which will give rise to the adult cuticular structures of head, thorax and genitalia), and as abdominal histoblast nests (which will give rise to adult abdominal structures). The stepwise determination of groups of imaginal disc cells during larval life, so that each group forms one particular element of the adult cuticle structure during metamorphosis, is discussed in section 7.3. *Transdetermination* is a change in the determined state of imaginal disc cells, resulting in the production of inappropriate adult structures from cells originally destined to form a different structure (§7.4).

During embryonic and larval development, each imaginal disc becomes progressively subdivided into *compartments* whose cells do not intermingle (§7.5). These distinctions result from the activities of *selector* genes, which act in a combinatorial fashion to specify the state of cellular determination (§7.4 and 7.6). Mutation of these genes results in *homoeotic* mutants, in which part or all of an adult cuticular structure (e.g. antenna) is transformed into another structure (e.g. second leg) characteristic of a different segment or compartment. Homoeotic mutations also affect the structure of larval segments, though not their number or size. Analogies with transdetermination are apparent in many homoeotic mutant phenotypes, as outlined in section 7.6.1. The selector genes defined by homoeotic mutations are the only clear examples yet identified in animals of genes whose activities direct whole pathways of development. However, the same basic cell types are involved in secreting adult cuticle structures in all the different segments and their compartments. Thus the identified selector genes act primarily on *pattern formation*, directing the extent and spatial organisation of cuticle secretion to form the hairs, bristles, plates and other regional characteristics of the larval and adult exoskeleton. To a large extent this also applies to the central nervous system and musculature (see end of §7.5).

7.2 Puffing

The giant polytene chromosomes in the salivary-gland nuclei of Dipteran larvae provide a unique opportunity for observing patterns of gene expression. One or more of the following features can be used to distinguish transcriptionally active bands from those which remain inactive.

(a) Radioactive uridine is incorporated into RNA at active but not inactive sites, as revealed by *in vivo* labelling followed by autoradiography.

(b) There may be gross changes in the state of DNA condensation at active sites, giving rise to expanded regions known as *puffs*. These are formed from the condensed chromomeric DNA of a band by a process analogous to untwisting the strands of a rope (fig. 64*A*; see Beerman, 1964). Such puffs are sites of active RNA synthesis, as defined by criterion (a) above. However, RNA synthesis also occurs at some non-puffed band sites (see Bonner & Pardue, 1977a). Probably puffing represents a modification of the DNA organisation required for very high levels of transcriptional activity (cf. lampbrush chromosome loops; see §4.4).

(c) *In situ* hybridisation of the polytene chromosomes with specific labelled probes (cellular RNA species, cDNAs or cloned DNAs) can identify those active puffs which are engaged in synthesising the corresponding primary transcripts (fig. 64*B*i). In tissues *not* expressing the probe sequence as RNA, this approach can be used to map the gene(s) to one (or more) inactive band(s) on the polytene chromosomes (fig. 64*B*ii).

(d) RNA polymerase II (§3.3) is located on polytene chromosomes at active puff sites. However, fluorescent-tagged antibodies against this polymerase label many inactive interband regions as well as the puffs (Jamrich *et al.*, 1977a,b). By contrast, immunofluorescence with antibodies against RNA polymerase I labels only the nucleoli, but not puffs or other sites on the polytene chromosomes (fig. 64*C*; Jamrich *et al.*, 1977b).

Puffing patterns are both stage- and tissue-specific, as would be expected if the puffs represent gene sites engaged in particularly active transcription; several examples of this specificity are outlined in sections 7.2.1 and 7.2.2 below.

Fig. 64*A*, *B*, *C* Transcription in polytene chromosomes.

A

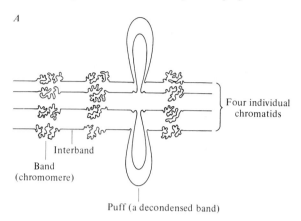

Four individual
chromatids

Interband

Band
(chromomere)

Puff (a decondensed band)

Fig. 64*A* Puffing (diagrammatic, after Beerman, 1964).

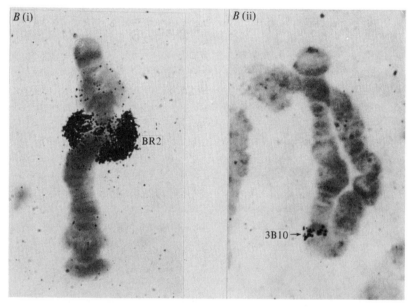

Fig. 64*B* *In situ* hybridisation of radioactive RNA from Balbiani ring 2 to *Chironomus* polytene chromosome IV. Photographs kindly supplied by Prof. B. Daneholt (Karolinska Institutet, Stockholm).
(i) Hybridisation to the active BR2 puff site on chromosome IV in a salivary gland nucleus. Reprinted with permission from B. Lambert (1972) *Journal of Molecular Biology* **72**, 65–75. Copyright 1972 by Academic Press Inc. (London) Ltd.
(ii) Hybridisation to the inactive 3B10 band on chromsome IV in a Malpighian tubule nucleus. From L. Wieslander and B. Daneholt, unpublished.

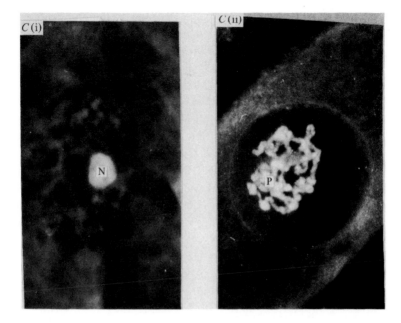

Fig. 64C Immunofluorescent localisation of RNA polymerases I and II
in *Drosophila* salivary gland polytene nuclei. Photographs kindly
supplied by Prof. E. K. F. Bautz (Molekular Genetik, University of
Heidelberg), from the thesis of M. Jamrich, University of Heidelberg,
1978. Used with permission.
(i) Nucleus stained with fluorescent-tagged antibodies against RNA
polymerase I. Staining essentially confined to nucleolus (N).
Magnification ×220.
(ii) Nucleus stained with fluorescent-tagged antibodies against RNA
polymerase II. Staining confined to the polytene chromosomes (P),
particularly at expanded puff sites. Magnification ×190.

7.2.1 *Balbiani ring 2 in* Chironomus

In the larval salivary glands of the midge *Chironomus*, several puffs attain
enormous sizes and are termed Balbiani rings (BRs). These are the sites
of synthesis for RNAs encoding a range of large polypeptide products
which are secreted in huge quantities by the larval salivary glands (see e.g.
Grossbach, 1973). Probably BR1 codes for the fraction 3 polypeptide,
while BR2 specifies the fraction 2 protein (Pankow *et al.*, 1976).

BR2 is the largest of these Balbiani rings in salivary gland cells, and is
located on the small chromosome IV (fig. 64*B*). Indeed, it is possible to
microdissect BR2 and so study its structure and RNA products in

isolation. In other larval cell types such as the Malpighian tubules (which also have giant banded chromosomes) this chromosomal site is not puffed, and *in situ* hybridisation with BR2 RNA identifies a single band in region 3B10 on chromosome IV (fig. 64*B*ii; Derksen *et al.*, 1980). This band contains some 470 kbp of DNA, far larger than the known size of the BR2 transcription unit (37 kb RNA product). Nevertheless, hybridisation data suggest no more than about four BR2 genes per DNA duplex in the 3B10 region, while EM studies detect only a single active BR2 transcription unit per chromatid (Lamb & Daneholt, 1979).

After spreading for EM, each BR2 transcription unit can be visualised as a Christmas-tree-like structure (fig. 65), reminiscent of the active ribosomal genes from *Xenopus* oocytes (§4.4). Very active transcription is implied by the close packing of polymerases and product chains. However, the ends of the RNA product chains begin to fold up into knobbed structures long before completion of their synthesis, probably through association with nuclear proteins to form RNP particles. Free RNP particles of the same type are abundant in the nucleoplasm of salivary gland nuclei, and in a few cases they have been observed apparently squeezing themselves through nuclear pores *en route* to the cytoplasm (Daneholt *et al.*, 1979).

The RNA synthesised at the BR2 site is somewhat peculiar in two respects. Firstly, it is extremely large (75S or 37 kb in length), and could potentially encode more than a single copy of the fraction 2 polypeptide. Secondly, some at least of this 75S RNA is transported intact into the cytoplasm, i.e. without major trimming or intron removal. Huge polysomes containing 60–100 ribosomes are found in *Chironomus* salivary-gland cytoplasm (Daneholt *et al.*, 1977; Francke *et al.*, 1982); these contain mRNA strands ranging up to 75S (37 kb) in size and complementary to BR2 DNA (Wieslander & Daneholt, 1977). Given the likelihood that some degradation will occur during the isolation of such long RNA molecules, this would suggest that the 75S BR2 transcripts undergo little if any size alteration within the nucleus. BR2 has been extensively reviewed as a model system demonstrating specific gene activity (e.g. Case & Daneholt, 1977, and Daneholt *et al.*, 1979), though it is perhaps atypical in some respects.

7.2.2 *Puffs responsive to ecdysone*

During the late third larval instar in *Drosophila*, a group of new puffs appears in salivary gland nuclei in response to rising levels of ecdysone. As mentioned above, this hormone initiates metamorphosis only in the absence of juvenile hormone. Only a few puffs (e.g. 68C) are active

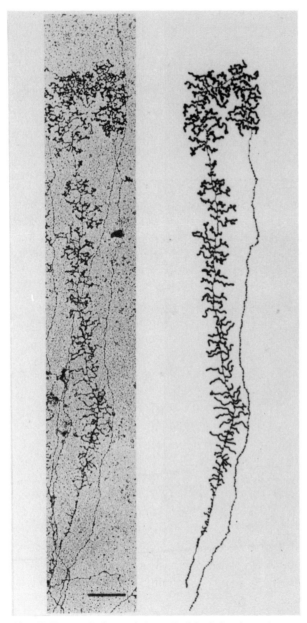

Fig. 65 Transcription unit from Balbiani ring 2 on chromosome IV in a *Chironomus* salivary gland nucleus. Electron micrograph of a spread preparation, by courtesy of Prof. B. Daneholt (Karolinska Institutet, Stockholm). Reprinted with permission from the copyright holders, *Cell*. From M. M. Lamb & B. Daneholt (1979) *Cell* **17**, 835–48. Note side-branches (representing nascent 75S RNA transcripts) folding into knobbed RNP structures as transcription proceeds. Bar at foot of electron micrograph (left) represents 0.5 μm.

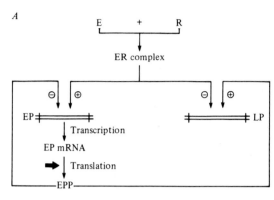

A

E   +   R
ER complex

EP ⊖ ⊕   ⊖ ⊕ LP
Transcription
EP mRNA
Translation ➡
EPP

E   Ecdysone
R   Receptor protein specific
     for ecdysone
EP   Early puff site(s)
LP   Late puff site(s)
EPP Protein product(s)
     from early puff(s)
⊕   Activation of puffing
⊖   Inhibition of puffing
     (i.e. puff regression)
➡   Blockage by cycloheximide

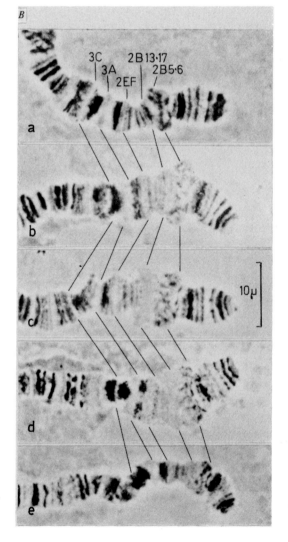

during the intermoult period, and these mostly regress in the presence of ecdysone.

The new puffs induced by ecdysone fall into two distinct classes (see Ashburner *et al.*, 1973). One group – termed the early puffs (such as 23E, 74EF and 75B) – begin to appear within minutes of hormone administration, reach their maximum sizes within 1–4 hours, and then regress. By contrast, the late puffs (such as 22C, 63E and 82F) appear only after a lag of about 3 hours and reach their maximum sizes some 8 to 10 hours after hormone administration, before eventually regressing. Inhibitors of protein synthesis such as cycloheximide block induction of the late puffs, implying that one or more proteins specified by the early puffs are responsible for activating the late puffs. Cycloheximide does not block the appearance of early puffs, so these must be directly induced by ecdysone itself. Moreover, two of them (74EF and 75B) fail to regress in the presence of cycloheximide, which further suggests that these early puffs are normally switched off by their own protein products. If ecdysone is washed out during the first four hours of treatment, then several late puffs are prematurely induced, implying that ecdysone itself may normally delay puff formation at these sites. The overall system of checks and balances which emerges from this study (Ashburner *et al.*, 1973) is summarised in fig. 66.

More recently, Walker & Ashburner (1981) have confirmed the main features of this model by using aneuploid *Drosophila* genotypes in which the chromosomal segments bearing two of the early puffs are either increased or reduced in dosage. When extra copies of these early puffs are available, the response of some (but not all) late puffs is greater and more rapid; moreover the early puffs are active for a shorter period. The converse is true when using genotypes with reduced numbers of early puff sites. Here the period of early puff activity is prolonged, and the response

Fig. 66*A* Model for ecdysone induction of early and late puffs (modified from Ashburner *et al.*, 1973).

B Puffing patterns in *Drosophila* salivary-gland polytene chromosomes following ecdysone treatment in culture. Photographs by courtesy of Dr M. Ashburner (Dept. of Genetics, University of Cambridge) and reprinted with permission from the copyright holders, Springer-Verlag. From M. Ashburner (1972) *Chromosoma* **38**, 255–81. This series of photographs shows the tip of the X chromosome; (a) control (no ecdysone); (b), (c), (d) and (e) after 1, 2, 8 and 12 hours of culture in the presence of ecdysone. An intermoult puff at site 3C regresses in culture (independent of ecdysone), while band 2B5-6 forms a small early puff in response to the hormone. This puff has regressed by 12 hr (part (e)).

of the same set of late puffs is reduced. These observations confirm the dual role of the early puff products in repressing their own synthesis and in activating certain late puff sites.

Bonner & Pardue (1977b) used *in situ* hybridisation to show that transcripts from the early puffs 74EF and 75B appear in the RNA population of salivary gland cells only after ecdysone treatment. Thus puff induction genuinely reflects the switching-on of transcription at specific gene sites within the puffed region. Among the new RNA species induced in response to ecdysone is the messenger coding for dopa-decar-boxylase (DDC), an enzyme involved in cuticle cross-linking and pigment production in hypodermal cells (Kraminsky *et al.*, 1980). Recently, a cloned DDC gene plus 5'-flanking sequences has been inserted into a transposable P element (see §1.10; O'Hare & Rubin, 1983) and then reintroduced by microinjection into early *Drosophila* embryos carrying a homozygous DDC mutation. Although the P element plus DDC gene region becomes integrated into the host chromosomes at a variety of sites different from the usual location of this gene, Scholnick *et al.* (1983) found that the introduced wild-type DDC gene is expressed under normal developmental control in terms of tissue- and stage-specificity. This powerful technique (see Flavell, 1983) should allow the regulatory mechanisms governing specific gene expression in *Drosophila* to be analysed in much greater detail than has hitherto been possible. In particular, deletion of parts of the 5'-flanking sequences should permit the identification of regions required for normal developmental control of the DDC (and other) genes.

As mentioned earlier, several intermoult puffs active in salivary-gland cells during the third larval instar regress in response to ecdysone. These include the 68C puff of chromosome 3, which contains a gene (*sgs* 3) encoding one of the larval glue proteins secreted by salivary-gland cells (cf. the Balbiani ring products in *Chironomus*). The molecular organisa-tion of some 50 kbp of DNA from the 68C locus has been analysed by Meyerowitz & Hogness (1982). The only transcribed genes present in this region are clustered together within a 5 kbp stretch of 68C DNA. Apart from the *sgs* 3 gene itself (giving a 1.1 kb RNA species), there are also two oppositely oriented genes with adjacent promoter sites, whose RNA products (0.32 and 0.36 kb) encode the newly identified glue proteins *sgs* 8 and *sgs* 7 (Crowley *et al.*, 1983). Thus three genes are located within a single band at the 68C locus.

There is now good evidence for a direct chromosomal action of ecdysone on puffing patterns (reviewed by Ashburner, 1980). This hor-mone's structure includes an α-β unsaturated ketone grouping which makes it possible to cross-link the molecule to its binding site by means of photoactivation. The irreversibly bound hormone can then be localised on polytene chromosomes by immunofluorescence using antibodies specific for ecdysone. This technique was used by Gronemeyer & Pongs (1980) to show that the hormone is bound specifically at ecdysone-inducible early puffs (see above), and also at intermoult puff sites such as 68C which regress in response to ecdysone. However, specific hormone binding does not completely explain the puffing changes observed in response to ecdysone (see Dworniczak et al., 1983). The same technique has been used recently to identify the ecdysone receptor (Schaltmann & Pongs, 1982). It is a cellular protein of MW 130000, and is found both in salivary glands and in *Drosophila* tissue-culture cells. It shows all the characteristics typical of hormone receptor proteins, including high-affinity binding of ecdysone and rapid accumulation of the hormone within the nucleus following ecdysone administration. As described above, the hormone binds specifically within the nucleus to many chromosomal loci whose expression is directly regulated by ecdysone. Essentially this is the same phenomenon as that described in the previous chapter for the hormonal induction of egg-white protein and vitellogenin gene expression.

7.3 Imaginal discs

As outlined in the introduction (§7.1), much of the adult or *imago* is built up during metamorphosis from reservoirs of imaginal cells which have remained diploid throughout larval life (reviewed in Ursprung & Nothiger, 1972). The discussion below will be confined to the imaginal discs, which because of their discrete nature and large size are much easier to study than the abdominal histoblast nests (see §7.1). Each imaginal disc comprises a single layer of ectodermal epithelial cells that will secrete a defined part of the adult cuticle during metamorphosis. Underlying this layer are the adepithelial cells, including myoblasts which probably form the adult musculature in the region specified by the disc.

The nineteen imaginal discs in *Drosophila* larvae comprise nine pairs of head and thoracic discs, plus a single fused abdominai disc which gives rise to the genital/anal structures of the adult. Three paired discs give rise respectively to the labial, clypeolabral and eye-antennal structures of the adult head, while six paired discs give rise respectively to the dorsal prothorax, wings, halteres, first, second and third legs of the adult thorax.

As implied by this classification, each disc is normally only capable of forming a limited portion of the adult exoskeleton, and is thus *determined* to follow a certain path of development during metamorphosis. In the late larval stage, this is also true for small groups of cells and even for some single cells within a given disc, each group of cells giving rise to one particular element of the adult structure (e.g. the different proximodistal regions of a leg). The stability of the determined state in imaginal disc cells when explanted into organ culture, will be discussed later in section 7.4.

A mature imaginal disc behaves as a *mosaic* of rigidly determined and largely autonomous cell regions. Usually damage to one part of the disc will result in the absence or malformation of the corresponding adult element after metamorphosis. However, if a sufficient period of growth elapses between the time of damage and metamorphosis, then either (i) the deficiency will be made good by regeneration, or else (ii) reduplication will occur (see Schubiger, 1971; Bryant, 1971).

Among the different types of imaginal disc cells during larval life, there are few if any indications of their future differentiation into specific adult structures. Morphologically, these cells are mostly small and of cuboidal epithelial type; there are no obvious distinctions in terms of fine structure between cells from different discs, let alone between cells from different regions of a single disc (see Ursprung, 1972). Only in the eye disc are any groups of cells clearly differentiated at the end of larval development. Thus the imaginal cells remain largely undifferentiated during larval life, and express their predetermined fates only in response to ecdysone (in the absence of juvenile hormone). The protein populations of imaginal disc cells are qualitatively and quantitatively very similar. When the patterns of newly synthesised or accumulated proteins are compared between different disc-types using 2D gels, very few consistent differences can be identified (Greenberg & Adler, 1982; Ghysen *et al.*, 1982). One protein is synthesised non-uniformly in different regions of both the wing and haltere discs (Greenberg & Adler, 1982); three muscle-specific proteins, contributed presumably by adepithelial myoblast cells, are accumulated to a greater extent in haltere and third-leg discs than in wing and second-leg discs (Ghysen *et al.*, 1982).

In the remainder of this section, I shall discuss the origin and stepwise determination of imaginal disc cells. One technique of key importance in elucidating this process is the genetic marking of *clones* of cells. When embryos or larvae are exposed to moderate doses of X-rays (around 1000 roentgens), somatic recombination is induced in a very small proportion of cells. (Normally recombination is confined to meiosis in the germ cells, but rarely it also occurs during mitosis in somatic cells; both processes involve crossing-over between homologous chromosomes). One can take advantage of this if the animals to be irradiated are heterozygous for a suitable recessive mutation, i.e. one which in homozygotes alters the adult cuticle characteristics in a cell-autonomous manner. In such a situation, somatic crossing-over during mitosis will sometimes result in one double-dominant and one double-recessive daughter cell (fig. 67). These cells will be respectively wild-type (like the heterozygous background) and mutant in character. Thus descendants of the latter will form a *marked clone* in the adult structure; that is, a region of cells distinguished by the mutant cuticle character (e.g. altered colour or presence of abnormal hairs). The principle is illustrated in fig. 67 for a recessive gene

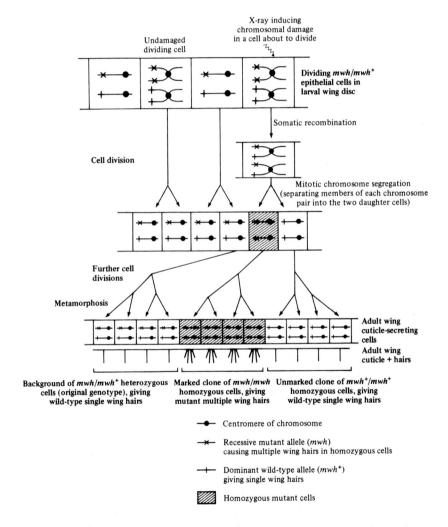

Fig. 67 Marking cell clones by means of X-ray-induced somatic crossing-over (modified from Garcia-Bellido *et al.*, 1979).

(*mwh*) which when homozygous causes multiple rather than single hairs to appear on the wing blade.

Cells destined to form the imaginal discs are set aside in the embryonic blastoderm, and thereafter remain separate from the cells forming larval structures. Available evidence from clonal marking experiments suggests that there are several founding cells per disc, and this feature also characterises the series of determination events which arise subsequently within each disc (see also section 7.5). Thus each determination decision is made by a *group* of cells which together give rise to a *polyclone*, as opposed to a clone produced from a single founding cell.

The evidence for this statement is derived from experiments where marked clones (fig. 67) are induced among dividing imaginal disc cells at successively later stages during development. Let us consider the case of a binary determination decision involving a choice between two alternative fates X and Y. Needless to say, clones induced early in development are larger than those induced later, since the original marked cells will have undergone more division cycles in the former case. If a large clone is induced several cell divisions before the decision to form X or Y, then regions of mutant cuticle may appear in both final structures; in other words, some of the marked progeny decide to form X while others follow the Y pathway. If a smaller clone is induced at a later stage in one of the cells already committed to form X, then a patch of mutant cuticle will appear only in structure X. But what happens when a marked clone is induced immediately prior to the X/Y determination decision? Two different results would be predicted, depending on whether the determination decision affects only one founding cell or several together. In the former case, it should sometimes be possible to mark the single founding cell so that the *entire* cuticle of structure X shows the mutant character. In the latter case, this would never be observed, since the rarity of somatic recombination events ensures that several founding cells could not become marked simultaneously. In fact the latter turns out to be true; in no case do marked cells form a complete adult structure (see fig. 68), thus several founding cells must take each determination decision *together* as a group.

Final commitment to form a particular element of the adult structure is not achieved all at once, but rather proceeds stepwise from the general (e.g. wing versus leg) towards the particular (e.g. distal wing versus proximal notum). An example of this will be given in section 7.5 (fig. 74), showing how imaginal cells become determined in several steps to form the various compartments of the second thoracic (wing and leg) discs. At each step a binary determination decision is apparently made, so that

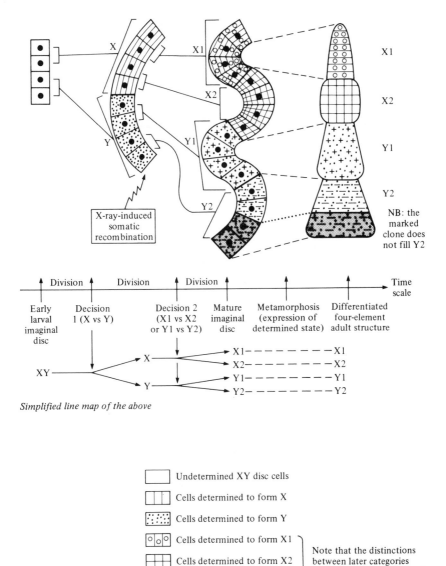

Simplified line map of the above

Fig. 68 Binary determination decisions in dividing imaginal disc cells (schematic).

groups of cells progressively acquire a detailed 'address' appropriate to their position and specifying their future fate. Evidence for this is again derived from marked clones induced at different stages in development. A clone induced in an early disc may give rise to a large mutant cuticle patch occupying parts of several elements in the adult structure produced by that disc (e.g. in both wing and notum). By contrast, a clone induced at a later stage will mark only one specific element in the final structure (e.g. wing *or* notum, but not both). In the example quoted, a decision to form distal wing or proximal notum must have occurred at some point between the times at which the two clones were induced. From a large series of marked clones induced at different times, one can deduce the timetable of successive determination decisions within a given disc-type (see figs. 68 and 74). These decisions occur hierarchically, i.e. in a specified order at defined stages of disc development.

Initially, only a small number of blastoderm cells (perhaps ten or so) is set aside to form each imaginal disc. By the time of pupation, many thousands of cells are present in each disc, e.g. about 50000 in a wing disc. Thus on average each founding cell must pass through some ten to twelve division cycles. However, mitotic rates are not uniform throughout the disc; rather, polyclonal groups of cells divide at their own characteristic rates, with some groups dividing faster or ceasing division earlier than other groups. Programmed cell death may also occur in particular regions of the disc. In its early stages each imaginal disc forms a simple inpocketing of the larval epithelial cell layer. As differential growth proceeds in different regions of the disc, so this structure becomes highly convoluted, giving complex shapes and folding patterns which are characteristic for each type of mature disc (fig. 69). Throughout all this, the ectodermal cells which will secrete the adult cuticle remain organised as a coherent monolayer i.e. a two-dimensional sheet of cells.

Particular regions of the mature disc are normally determined to form discrete elements in the final adult structure. This has enabled two-dimensional 'fate maps' to be derived for each type of disc, showing which groups of disc cells produce which adult elements. Each such map reflects the hierarchical series of determination decisions taken by groups of cells within the growing disc (fig. 68). During the final stages of imaginal disc development, determination decisions may also be taken by individual cells, so that one particular cell becomes destined to produce say a single bristle in the adult structure. However, the process of cellular determination often stops at a so-called 'prefinal' state, leaving individual cells with a very limited range of developmental options still open to them at the onset of metamorphosis. The final choice between these options has been

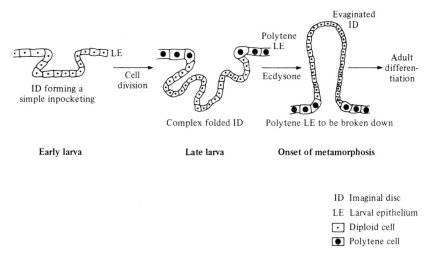

Fig. 69 Imaginal disc development (diagrammatic).

shown in some cases to depend on inductive interactions between neighbouring cells; thus special trichomes called bracts are only formed at the base of leg chaete organs, and never in isolation (see Nothiger, 1972).

During metamorphosis, ecdysone induces a change in cell shape (from cuboidal to squamous epithelial) among the imaginal disc cells. This causes each disc to become everted like a balloon (fig. 69), from which the final adult structure is modelled. For instance, flattening occurs during wing formation to produce an ovoid double layer of cells.

7.4 Transdetermination: the stability of the determined state

One advantage of studying imaginal discs (as opposed to abdominal histoblast nests) is that they can be readily isolated from the surrounding larval tissues; complete discs or parts thereof can then be cultured for prolonged periods *in vitro* or *in vivo*. In the former case, disc explants are maintained in a suitable culture medium where the cells may continue to divide but do not differentiate into adult structures unless ecdysone is added (Edwards *et al.*, 1978; Seybold & Sullivan, 1978). *In vivo* culture methods involve transplanting isolated discs or disc fragments into the abdomens of host larvae or adults. Normally one would choose a host strain carrying different genetic markers from the donor strain, so that any invasion of the transplant by host cells could be readily detected; in practice, however, this does not seem to occur, and the transplant develops autonomously. In a larval host, the transplanted disc material will differentiate into adult structures only when high ecdysone levels are encountered, i.e. at the time of host metamorphosis. In the haemolymph of adult hosts, however, the concentration of ecdysone is extremely low, since high hormone levels are required only to initiate metamorphosis and not to maintain the adult state of differentiation. Thus transplants of disc material will continue to grow but will never differentiate in the environment of an adult host abdomen. Such disc transplants can be transferred repeatedly from one adult abdomen into another (the life span of adult flies being limited); this sequence of *serial transplants* can be extended over many *transfer generations* lasting months or even years (Hadorn, 1978). Only after retransplantation into a larval host can the cultured material undergo adult differentiation, i.e. at the time of host metamorphosis.

The state of determination is normally very stable in imaginal disc cells (see Gehring, 1972), and is passed on unchanged from parent to daughter cells (i.e. it is a heritable characteristic; see later). As mentioned above, disc explants can be cultured for prolonged periods in the absence of ecdysone without ever differentiating into adult structures. However, when exposure to ecdysone eventually occurs (*in vitro* or in a metamorphosing larval host), the predetermined pattern of adult differentiation is usually expressed. Thus cells derived from part of a leg disc will give rise to the appropriate leg elements which those cells would have produced if left to metamorphose *in situ*. If one allows an excessive number of cell divisions to occur before treating with ecdysone, then the abnormally large mass of disc tissue which results will usually produce two or more reduplicated but normal-sized elements of the adult structure, rather than a single outsize element. In some instances other elements of the same adult structure may be produced by regeneration (Schubiger, 1971; Bryant, 1971).

Occasionally, however, this stable inheritance of the determined state breaks down. After prolonged culture, cells derived from say a leg disc may give rise to adult structures clearly recognisable as wing or (rarely) antennal elements. These inappropriate or *allotypic* structures are said to be *transdetermined* (Hadorn, 1978), since they apparently result from a change in the state of determination among imaginal disc cells. A greater or lesser proportion of the final adult structure may show the transdetermined phenotype, depending on when the transdetermination event occurred (i.e. how many cell divisions have elapsed since then). This generally gives an adult structure in which a patch of tissue expressing allotypic characteristics is present together with tissue expressing the normal determined phenotype (*autotypic* characteristics). The two are

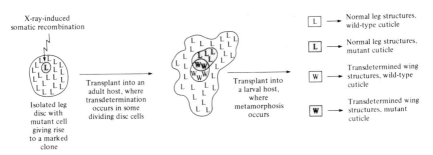

Fig. 70 Groups of cells are involved in transdetermination (modified from Alberts *et al.*, 1983).

not intermingled and the boundaries between them are sharp, i.e. with no intermediate regions sharing say leg and wing characters. Clonal marking experiments (see §7.3) confirm that transdetermination, like normal determination, occurs in groups of cells. Thus marked clones induced prior to transdetermination may give rise to part but never to all of the region manifesting the transdetermined phenotype (fig. 70).

Transdetermination changes the determined state in a group of imaginal disc cells so that their progeny follow a developmental pathway different from that to which they would normally have been committed. Though inappropriate for cells belonging to the original disc-type, this new pathway is one which would be perfectly normal for cells from a different disc-type, or from a different region of the same disc (Hadorn, 1978). Only a limited choice of transdetermination options is available to the cells from any given disc-type; some types of transdetermination events are far more frequent than others (fig. 71), while some are apparently impossible (i.e. have never been observed to occur). Moreover, cells which are already transdetermined may sometimes undergo further transdeterminations, or sometimes revert to their original determined state. This can be shown by serial transplant experiments in which part of the disc material from each transfer generation is placed in a larval host and allowed to metamorphose, while the rest is returned to a new adult host (Hadorn, 1978).

These features suggest that we are not dealing with a classical mutational event, which would be expected to disrupt development far more often than merely switching pathways. Also, the frequency of transdetermination is far greater than could be explained by spontaneous mutations, and known chemical mutagens do not increase this frequency.

Nevertheless, both the autotypic and allotypic determined states are heritable, being passed from one original group of cells to their progeny.

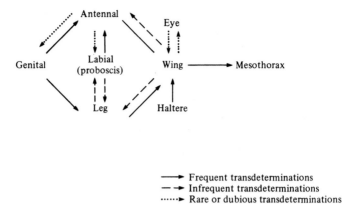

——————▶ Frequent transdeterminations
— — ▶ Infrequent transdeterminations
·······▶ Rare or dubious transdeterminations

Fig. 71 Frequency of transdetermination events (after Hadorn, 1978).

One likely explanation is that a pattern of gene activity can be inherited from parent to daughter cells. If so, the limited number of available options would imply that determination decisions involve activating and/or inactivating a fairly small number of regulatory genes. Changes in this pattern would result in discrete alterations to the determined state, as observed in transdetermination. If determination involves binary choices (e.g. between 'off' and 'on' states for the activity of a particular regulatory gene), and if n different regulatory genes are involved in specifying the final determined state, then there are 2^n combinatorial possibilities for patterns of regulatory gene activity, each of which could represent a different determined state. There is indeed evidence for such regulatory genes (*selector* genes, Garcia-Bellido, 1975; see §7.6), and for the idea that a final determined state is arrived at combinatorially through a sequence of binary determination decisions (see figs. 68 and 74). This model is probably oversimplified, but shows how a few regulatory genes might be used very economically to control many developmental pathways. Thus only four such genes would be needed (in all possible combinations of 'off' or 'on' states) to specify 16, i.e. 2^4, different determined states.

There is a clear correlation between cell-division rates and transdetermination frequency, as discussed by Hadorn (1978). Thus labial disc explants which are not permitted to undergo extra growth during explant culture do not give transdetermined structures. Male genital discs remain slow-growing during the first few transfer generations, and at this stage they very rarely transdetermine; subsequently, however, they enter a rapid growth phase when transdetermination occurs much more frequently. This conclusion is confirmed by experiments manipulating the growth rate of disc cells in explant culture; allotypic structures arise far more frequently from fast-growing than from slow-growing explants. Similarly, prolonged culture periods prior to metamorphosis increase the likelihood of transdetermination, provided that some cell division is occurring. Overall, only dividing cells have been observed to transdetermine, possibly because the pattern of gene activity can only be reprogrammed during mitosis (the same would also presumably apply to normal determination decisions). High division rates would therefore favour transdetermination if at each mitosis there is a small chance of changing the preestablished state of determination into a new one (perhaps by affecting the activity of a single regulatory gene; see above).

7.5 Segments and compartments in insect development

The insect body is characteristically subdivided into segments. In *Drosophila*, several segments (possibly six) are fused to form the head, while the thorax is composed of three segments (each carrying a pair of legs), and the abdomen of eight segments (plus a latent ninth segment at the posterior end). In primitive segmented creatures such as annelid worms and myriapods (millipedes etc.), most of the segments are very similar in structure, with significant modifications mostly confined to the anterior end. Embryogenesis in such animals involves a series of reiterated developmental units each corresponding to one segment. There is evidence that similar organisational principles underlie insect development, though somewhat obscured by the superimposed differences between segments. As mentioned in the introduction (§7.1), the multilayered germ band (formed by folding the blastoderm) soon becomes subdivided into segments, and this embryonic pattern of organisation persists through the larval stage and is reflected in the adult (fig. 72). Indeed, the one-to-one correspondence of embryonic, larval and adult segment patterns allows one to predict the likely adult phenotypes of embryonic-lethal homoeotic mutants (see §7.6 below) from the cuticular patterning of late embryos which are homozygous for the mutant character.

During insect development, cells from one segment do not trespass into another. Thus each segment constitutes a separate developmental *compartment*, which is *polyclonal* because it is derived from several founding cells (see review by Lawrence, 1981a). In the blastoderm each segment is founded by two discrete groups of cells, one forming the right half and one the left (Garcia-Bellido & Merriam, 1969). As mentioned above (§7.3), the imaginal discs are all bilaterally paired except for the single genital/anal disc in which the right and left halves are fused. These

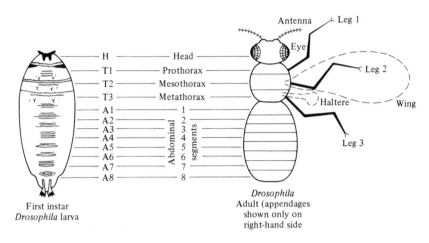

Fig. 72 Segment patterns of larval and adult *Drosophila*.

discs must also be organised in a basically segmental pattern, although there is no simple one-to-one correspondence between discs and segments. Nevertheless, as shown in fig. 72 the overall number and arrangement of segments is the same in adults as in larvae. Imaginal discs, like segments, are discrete development units. If cells belonging to different disc types are dissociated and mixed together *in vitro*, they do not intermingle, but rather sort themselves out in a disc-specific manner reflecting different cellular adhesion affinities.

The segment pattern in *Drosophila* is controlled by a relatively small number of genes. Nüsslein-Volhard & Wieschaus (1980) have identified fifteen loci which, when mutated, alter the larval segmentation pattern (these mutants are all late embryonic lethals). Mutations at six of these loci cause pattern duplications within each segment (segment polarity mutants), six cause pattern deletions in every other segment (pair rule mutants), and three cause deletions of a group of adjacent segments (gap mutants). Homozygous embryos carrying mutations at the pair-rule locus *fushi tarazu* (*ftz*) have only half the usual number of segments, the posterior half of every even-numbered segment and anterior half of every odd-numbered segment being deleted. *In situ* hybridisation reveals that *ftz⁺* transcripts are normally expressed early in embryonic development; by the cellular blastoderm stage, they become confined to seven evenly-spaced bands, corresponding roughly to those regions deleted in *ftz*/*ftz* mutant embryos (Hafen *et al.*, 1984).

Within each developing segment or imaginal disc, discrete compartments arise – distinguishing anterior from posterior, dorsal from ventral and sometimes proximal from distal groups of cells. These categories result from a hierarchical series of determination decisions taken, in the above order, by several cells together. The proof that each compartment has a polyclonal origin is essentially the same as that outlined in section 7.3; i.e. by generating marked clones just prior to the time of determination and showing that marked cells never completely fill a compartment, which implies that there must be more than one founding cell.

Operationally, cells belonging to neighbouring compartments in a disc never cross the compartment boundary during development *in situ*. This is important because the borderlines between different compartments do not always correspond to structural discontinuities apparent in the imaginal disc or adult cuticle. In the case of wing development, for instance, the boundary between anterior and posterior compartments does not run along the line of any wing vein (fig. 73). Nevertheless, clonal marking experiments clearly demonstrate the existence of an otherwise-invisible frontier between anterior and posterior compartments in the wing disc (see Crick & Lawrence, 1975). Patches of mutant cuticle derived from marked clones normally have irregular outlines, but wherever one abuts on the compartment border that edge is unusually smooth and straight (fig. 73). However, each marked clone usually contributes only a small region of cuticle to the final adult structure, hence a large number of instances is required in order to establish the position of a given compartment boundary.

A more dramatic demonstration of compartment borders can be made in *Drosophila* heterozygotes carrying the dominant *Minute* (M) mutation, which are characterised by an abnormally slow rate of cell division. If this strain is also heterozygous for a closely linked recessive cuticle marker such as *mwh* (*multiple wing hairs*; see fig. 67) in the combination M^+ *mwh/M mwh$^+$*, then X-ray-induced somatic recombination will sometimes generate clones of M^+*mwh/M$^+$mwh* homozygous cells. These clones will be *fast*-growing (M^+/M^+), and will give rise in the adult to multiple hairs on the wing blade (*mwh/mwh*). The surrounding cells will be M^+*mwh/M mwh$^+$* (original genotype) and will thus grow slowly, giving rise to wild-type single hairs on the wing (M and *mwh$^+$* being dominant to M^+ and *mwh*). Since the M/M genotype is lethal, there will be no contribution from M *mwh$^+$/M mwh$^+$* cells (i.e. the other homozygous genotype generated by somatic recombination). Thus the overall result of this genetic trick (Garcia-Bellido *et al.*, 1973; Morata & Ripoll, 1975) is to generate fast-growing clones of marked cells on a slow-growing unmarked background. Such a clone may nearly fill a compartment, though never completely so since several cells are involved in founding each compartment (all the others will be slow-growing; see fig. 73). Most importantly, such *fast-growing clones never transgress the compartment boundaries* once established (fig. 73), nor is the final size of the compartment altered (Garcia-Bellido *et al.*, 1973; Crick & Lawrence, 1975). This implies that cell division is halted even in fast-growing clones once the compartment has reached its preset size-limit. This regulation is not achieved by counting cell divisions within clones of cells, otherwise fast-growing clones would not outgrow slow-growing clones (the latter passing through fewer division cycles than usual and the former through more). Presumably spatial cues exist within the developing compartment, to which cells respond by ceasing division once it is filled.

A   X-ray induction of marked *mwh/mwh* clones in posterior compartments of wing discs

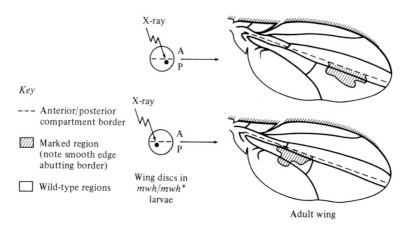

Key

--- Anterior/posterior
    compartment border

▨ Marked region
    (note smooth edge
    abutting border)

☐ Wild-type regions

X-ray

Wing discs in
mwh/mwh⁺
larvae

Adult wing

B   X-ray induction of marked fast-growing *M⁺ mwh/M⁺ mwh* clone in posterior compartment of wing disc

Key

--- Anterior/posterior
    compartment border

☐ Marked region from
    fast-growing clone
    (almost but not quite
    filling posterior
    compartment)

☐ Wild-type regions

X-ray

Wing discs in
M mwh⁺/M⁺ mwh
larvae

Adult wing

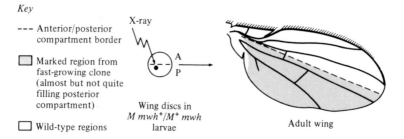

Fig. 73 Anterior and posterior compartments in the *Drosophila* wing disc (after Crick & Lawrence, 1975).

As mentioned earlier, the distinction between anterior and posterior compartments is only the first in a series of determination decisions which arise within the wing disc (see Morata & Lawrence, 1977). During the later development of this disc, a distinction is made between the dorsal and ventral compartments which will form, for example, the upper and lower surfaces of the wing respectively. Later still, a proximal/distal boundary is established, distinguishing those cells which will form the proximal notum from those which will form the distal wing-blade. In fact, the posterior/anterior distinction is apparent as soon as the appropriate segmental compartment (2nd thoracic) is established in the blastoderm. At this early stage the presumptive wing and second-leg imaginal cells are still grouped together, though they split to form separate imaginal discs soon afterwards (fig. 74). Thus it is possible to induce early clones which will give rise to marked structures in the anterior but not posterior regions of both wing and second leg, or else in the posterior but not anterior regions of both. If induced slightly later in development, a marked clone would affect only the wing *or* the leg cuticle, but not both together. Later again, a marked region might arise only on the dorsal but not ventral wing surface, or *vice versa*. Still later, marked structures might be confined to the proximal notum *or* to the distal wing-blade, but not both. During this hierarchical series of determination events (each of which appears to be binary), the previously established compartments are always respected so that each group of cells acquires, step by step, an increasingly detailed 'address' (shown schematically in fig. 74).

The borderlines between these eight compartments (fig. 74) in the mature wing disc are real in that, once established, they are never

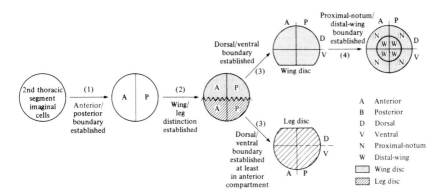

Fig. 74 Compartment determination among imaginal cells of the second thoracic segment (after Morata & Lawrence, 1977).

transgressed during normal development. In some cases the boundary is fairly obvious as a gross structural discontinuity; e.g. the wing margin separating dorsal and ventral compartments, or the proximodistal distinction between wing and notum. In other cases, however, the boundary corresponds to no obvious feature in either the imaginal disc or the final adult structure, as in the case of the anterior/posterior border running across the wing disc and wing (fig. 73). Wilcox *et al.* (1981) have shown that a cell-surface antigen, which is initially present on all *Drosophila* wing disc cells, becomes progressively confined to cells of the dorsal compartment during the last larval instar. Its final distribution is precisely contained by the dorsal/ventral boundary as defined by clonal marking experiments. This suggests that the disappearance of this antigen from the cell surface is a characteristic feature of ventral but not dorsal cells. Note, however, that this restriction occurs *after* the establishment of these compartments, and is thus a consequence rather than a cause of the dorsal/ventral distinction. The boundary between the dorsal and ventral compartments is also marked by a line of unusually shaped cells running across the wing disc (Brower *et al.*, 1982), though no such feature can be distinguished for the anterior/posterior border.

All of the above refers to features of the adult exoskeleton, i.e. the cuticle secreted by ectodermal epithelial cells. To what extent the compartment model also applies to internal structures remains less certain. There is now a suitable clonal marker for internal tissues, expressed *inter alia* in muscle cells; namely, temperature-sensitive mutations affecting the mitochondrial enzyme succinate dehydrogenase (whose activity can be readily detected by histochemistry; Lawrence, 1981b). Using this marker, Lawrence & Brower (1982) have shown that implants of adepithelial myoblast cells from the wing-disc region can contribute to muscles throughout the adult fly, even though these muscles form precise segmental and compartmental sets. At first sight, this seems to imply that the segmental pattern of adult muscles is specified by the overlying epithelium, and not by selector-gene activity within the myoblasts themselves. Nevertheless, selector genes in the bithorax complex (see §7.6) are expressed in the embryonic mesoderm of T3 and abdominal segments (Akam, 1983). Moreover, recent data from Lawrence & Johnston (1984) suggests that the segmental specification of muscle sets is independent of the overlying epithelium. For instance, one particular muscle in segment A5 (fig. 72) is present only in males; homoeotic mutants (§7.6 below) which transform A6 and A7 towards A5 result in extra copies of this muscle being produced in male but not in female adults. Similarly, a homoeotic mutation converting A4 towards A5 gives re-duplication of this same muscle in the A4 as well as A5 positions, again only in males. Using genetically marked male/female mosaics (gynandromorphs), it is possible to generate adults whose abdominal epithelium is entirely female while the underlying muscle tissue is male. In these cases, the number of A5 male-specific muscles produced in the adult is found to depend on homoeotic mutations carried only by the male cells in the mosaic (the female cells being wild-type). Thus the segmental specification of muscle cells is apparently autonomous and is not imposed by the overlying epidermis. The earlier result of Lawrence & Brower (1982) may perhaps result from reprogramming of the transplanted myoblast nuclei, since each transplanted myoblast must fuse with several other myoblast cells (presumably mainly of local origin) at the transplant site in order to produce a syncitial muscle fibre (Lawrence & Johnston, 1984; see also §4.8).

In the central nervous system (CNS) there is also some evidence that segmental characteristics are specified by selector-gene activity within the CNS itself. One example concerns the role of bithorax-complex genes in establishing differences between the thoracic and abdominal segmental ganglia of the adult ventral nerve cord (Teugels & Ghysen, 1983); the three thoracic ganglia are large and paired (having major nerve connexions to the legs on either side), while the abdominal ganglia are smaller and fused. Homoeotic mutations in the bithorax complex which cause transformations of A1 towards T3 (bxd^- mutants; §7.6 below), often also result in a fourth paired ganglion structure at the site normally occupied by the A1 fused ganglion. A converse transformation of T3 towards abdomen (*Hab* mutants) often converts the paired thoracic ganglion of T3 into an abdominal-type fused ganglion, so that only two paired ganglia remain (in T1 and T2). Using weak alleles of these homoeotic mutations, the ganglion changes are sometimes found to occur in the absence of leg changes (and *vice versa*), indicating that thoracic-type ganglia are not dependent on nerve connexions to a leg, but on selector gene activity within the nerve cord itself.

7.6 Homoeotic mutants and selector genes

7.6.1 Analogies between homoeotic mutants and transdetermination

Homoeotic mutations result in bizarre alterations to the body plan of *Drosophila* and other insects. Essentially these involve a transformation of the structures normally present in one body segment (or compartment) into those appropriate to a different segment (or compartment). Thus in the mutant *ophthalmoptera*, wings are found sprouting from the head in the place of eyes. Certain combinations of *bithorax* and *postbithorax* mutations (see below) produce a four-winged fly in which the third thoracic segment (T3, metathorax) is transformed into a replica of the second (T2, mesothorax); the result is that both segments carry wings and second legs, but the metathoracic halteres and third legs are absent. Several dominant mutations of the *Antennapedia* locus give second legs in the place of antennae, while recessive null alleles of this same locus result in the converse transformation – i.e. the second legs are replaced by antennal structures (Struhl, 1981a). It should be noted that the above descriptions refer to the most extreme manifestations of these homoeotic mutations; in many cases the transformations are only partial and interpretation is correspondingly more difficult.

Several similarities with transdetermination are apparent in these extreme homoeotic phenotypes. Firstly, both types of abnormality affect the state of determination among imaginal disc cells, though homoeotic mutations also cause transformations in the corresponding larval segments (see below). Secondly, the pattern of adult development expressed by a particular imaginal disc or compartment is changed from its normal course into an alternative pathway corresponding to that expected from a different disc or compartment. Finally, only a limited range of alternative options is available; many of the transformations encountered in transdetermination have direct counterparts in known homoeotic mutants and *vice versa*, though there are exceptions on both counts (i.e. cases where no such counterpart has been described).

Both phenomena must in some way involve the genes controlling the determination of segments or compartments, i.e. the process whereby cells acquire a specific 'address' (see §7.5 and fig. 74 above). Since transdetermination is not a mutational event (§7.4), we must rule out the possibility that spontaneous homoeotic mutations might arise in cultured imaginal disc cells during transdetermination. This is also excluded by the polyclonal origin of transdetermined structures (see §7.4), since similar mutations could not possibly occur in several cells together. More plausibly, transdetermination might affect the expression of certain *selector* genes, whose overall pattern of activity could specify the determined state in a combinatorial fashion (as discussed at the end of section 7.4). This model also suggests that an appropriate pattern of selector gene activity could become established *progressively* in a particular group of cells (compartment) during development, resulting in stepwise determination towards a restricted developmental fate (see also §7.5). The remainder of this section will discuss several classes of homoeotic mutations, and will show that the wild-type products of these 'homoeotic' genes fulfil the functions predicted for selector-gene products (Garcia-Bellido, 1975).

7.6.2 *The* engrailed *locus*

Homozygous mutations at the *engrailed* locus (*en/en*) show a characteristic alteration of the adult wing structure, such that the posterior region comes to resemble a mirror-image of the anterior region. This is apparent to a large extent in the pattern of wing venation, and also in the presence of bristles along the posterior margin of the wing (normally these are confined to the anterior margin), as shown in fig. 75. By X-irradiating larvae heterozygous for a recessive mutation at the *engrailed* locus, it is possible to obtain homozygous mutant clones expressing the *engrailed* mutant phenotype on a wild-type background. Moreover, large mutant clones may be generated in animals heterozygous for the *Minute* character, as described in section 7.5. This gives fast-growing *engrailed* clones on a slow-growing wild-type background.

Mutant *engrailed* clones induced in the anterior wing compartment appear completely normal; moreover, whatever their size, they always respect the anterior/posterior compartment boundary (fig. 75). This suggests that the wild-type product from the *engrailed* locus is not required in anterior cells. However, if such clones arise in the posterior compartment then the mutant phenotype of altered venation and posterior marginal bristles appears within the clone but not elsewhere (fig. 75). Moreover, such posterior mutant clones do not respect the normal compartment boundary, but rather spill over into regions normally occupied by anterior cells (fig. 75). This suggests that the wild-type product of the *engrailed* locus is required only in posterior cells, in order to effect the normal anterior/posterior distinction (Lawrence & Morata, 1976). Thus *engrailed* mutants apparently define a selector gene involved in the determination of posterior versus anterior compartments.

Nor is this effect of the *engrailed* locus confined to the wing disc of the second thoracic segment (T2). The structures of the second legs (also T2), halteres (T3), and abdominal segments (Kornberg, 1981) are also transformed in *engrailed* mutants, such that their posterior regions come to resemble mirror images of their respective anterior regions. The same is true for the posterior compartment in that region of the head specified by the eye/antennal disc (Morata *et al.*, 1983). Thus expression of the wild-type *engrailed* gene is probably necessary in the posterior compartments of all these segments in order to establish the normal anterior/posterior distinction (see also Lawrence & Struhl, 1982). This suggests that a

A   Phenotype of wings and halteres in wild-type and *engrailed* flies

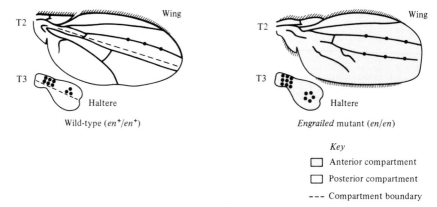

Wild-type (*en⁺/en⁺*)   *Engrailed* mutant (*en/en*)

Key

☐ Anterior compartment

☐ Posterior compartment

--- Compartment boundary

B   Fast-growing *engrailed* clones (*M⁺ en/M⁺ en*) induced in the wing discs of *M⁺ en/M en⁺* larvae

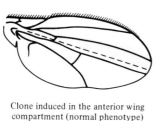

Clone induced in the anterior wing   Clone induced in the posterior compartment
compartment (normal phenotype)   (mutant phenotype in clonally marked
                                             region, which spills over into the anterior
                                             compartment)

☐ Marked fast-growing *engrailed* clone

☐ Slow-growing wild-type background

--- Anterior/posterior compartment boundary

C   Pattern of *engrailed* gene expression in T2 and posterior segments

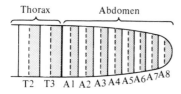

☐ *engrailed⁺* gene off (anterior compartment)

▨ *engrailed⁺* gene on (posterior compartment)

— Segment boundary

--- Anterior/posterior compartment boundary

Fig. 75 *Engrailed* gene function in *Drosophila* (based on Garcia-Bellido
et al., 1979).

'zebra-stripe' pattern of wild-type *engrailed* gene expression is superimposed on the segmental pattern, with this gene switched on in all the posterior compartments but off in all the anterior compartments (fig. 75). Recent molecular studies of the *engrailed* locus (Kornberg *et al.*, O'Farrell *et al.*, quoted in North, 1983) suggest that it spans some 60 kbp of DNA, and gives rise to several discrete transcripts in a pattern that changes during development.

7.6.3 The bithorax complex

A group of at least nine linked 'genes' forms the bithorax complex (BX-C), which governs the distinctions between abdominal and thoracic segments posterior to the mesothorax (T2). A simple model for the action of individual loci within this complex proposes that none of them are active in T2, but that more and more of them become activated in successively more posterior segments, until all are active in A8 (Lewis, 1978). This model, as we shall see, is in some respects oversimplified, but will serve as a basis for the discussion below. Individual homoetic mutations within the BX-C fall into two classes; (i) recessive loss-of-function mutants in which particular segments tend towards a more anterior condition, and (ii) dominant gain-of-function mutants which transform certain segments towards a more posterior condition (see Lewis, 1978). Examples of the latter group include the mutant *Contrabithorax* (*Cbx*), in which the mesothorax T2 acquires some characteristics of the metathorax T3 (e.g. a haltere-like transformation of the wings). Another example is the mutant *Hyperabdominal* (*Hab*), in which both the metathorax (T3) and first abdominal segment (A1) approach the condition of the second abdominal segment (A2).

Among the recessive loss-of-function mutants, perhaps the most dramatic example is provided by homozygotes in which the entire bithorax complex is deleted (denoted BX-C⁻). Not surprisingly this is a lethal mutation, but the embryo develops far enough to discern the larval segmentation pattern (fig. 76). The head, prothoracic (T1) and mesothoracic (T2) segments all appear normal, but the remaining segments (T3 and A1-8) are transformed to resemble T2 (fig. 76). However, A8 undergoes a partial transformation towards prothorax (T1), while a latent ninth abdominal segment is expressed without apparent conversion towards thorax. A reinterpretation of the phenotype of BX-C⁻ embryos has been suggested recently, whereby all segments from T2 to A8 develop identically, such that their anterior compartments are of T2-type and their posterior compartments of T1-type (Lawrence & Morata, 1983). This view is supported by evidence from a variety of mutants (see Morata &

Kerridge, 1981; Hayes *et al.*, 1984; Struhl, 1984a), all implying that the domains of action for particular BX-C genes are out of register with the segment pattern. Rather than specifying the anterior and posterior compartments of a single segment, individual genes in the BX-C may act within the posterior compartment of one segment and the anterior compartment of the next-most-posterior segment (hence e.g. posterior T2/anterior T3, or posterior T3/anterior A1). The original concept of in-register (segment-specific) domains of action stemmed from the relative paucity of larval cuticular markers in posterior as compared to anterior compartments. In the discussion below, 'T2-like' should be taken to mean 'posterior T1/anterior T2', and so on.

The effects of deleting the entire BX-C suggest that T2 (or rather anterior T2/posterior T1) is a kind of ground state, which is modified in successively more posterior segments by the activity of genes within the bithorax complex. Individual recessive mutants within this complex confirm this interpretation. Thus homozygous *bithorax* (*bx/bx*) mutants show a transformation of the anterior compartment of T3 towards that of T2; notably the anterior part of each haltere resembles an anterior wing region whereas the posterior part retains its haltere characteristics. Other anterior functions of the BX-C include *anterobithorax* (*abx*$^+$) required like *bx*$^+$ in anterior T3, and *postprothorax* (*ppx*$^+$; Morata & Kerridge, 1981) required in posterior T2 to distinguish it from posterior T1. Homozygous *postbithorax* (*pbx/pbx*) mutants change the posterior compartment of T3 towards that of T2, so that only the posterior region of each haltere acquires posterior wing characteristics. Deficiencies at all three loci required in T3 (*abx bx pbx/abx bx pbx*, or *abx bx pbx*/BX-C$^-$) notably give rise to four-winged flies, with T3 completely transformed into T2 (fig. 77). Homozygous *Ultrabithorax* (*Ubx/Ubx*) mutations transform both T3 and anterior A1 into T2-like structures (the posterior compartments of both T2 and T3 resembling posterior T1; Hayes *et al.*, 1984). In fact, *Ubx*$^-$ mutations inactivate the *ppx*, *abx*, *bx*, *pbx* and *bxd* functions (see below), suggesting that these loci are in some way subordinate to *Ubx*. The implication (above) that the true 'ground-state' is anterior T2/posterior T1 rather than simply T2, can be explained if some element of the *Ubx* 'domain' (*ppx*$^+$) is normally expressed in the posterior but not anterior compartment of T2, so distinguishing posterior T2 from posterior T1.

Similar transformations affecting abdominal segments involve recessive mutations of several further genes within the BX-C locus. In *bithoraxoid* (*bxd/bxd*) mutants, the anterior compartment of segment A1 is transformed towards that of T3, such that T3-type legs may appear on A1. Thus wild-type *bxd*$^+$ activity is normally required in anterior A1 but

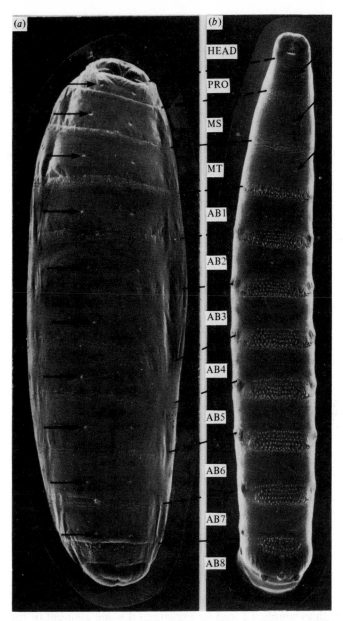

Fig. 76 Photographs of BX-C⁻ (a) and wild-type (b) *Drosophila*
embryos, reprinted by permission from Prof. E. B. Lewis (California
Institute of Technology, Pasadena) and from *Nature* **176**, 565–70.
Copyright © 1978, Macmillan Journals Ltd. Note the presence of
Keilin's organs (arrowed) and narrow denticle bands (typical of thorax)
around most of the thoracic and abdominal segments in (a), as compared
to the absence of Keilin's organs and wider denticle bands around the
abdominal (AB) segments in (b). Thus most of the abdominal segments
have been transformed towards a thoracic (T2-like) phenotype in the
absence of BX-C gene functions (a).

Fig. 77 Four-winged adult *Drosophila* showing complete conversion of segment T3 into T2 (i.e. carrying a second pair of wings in place of halteres); this mutant is deficient in the wild-type products of the *bx*, *abx* and *pbx* sites within the bithorax complex (see text). Photograph kindly supplied by Prof. E. B. Lewis (California Institute of Technology, Pasadena). Unpublished, from Biology Annual Report 1983 Cal. Tech.; permission granted for use.

not in anterior T3 nor in T2. The products of the wild-type *infra-abdominal* genes (e.g. $iab2^+$, 5^+, 8^+) are likewise involved in specifying more posterior abdominal segments; for instance, $iab2^+$ gene activity distinguishes A2 from A1, and the absence of this product converts A2 towards A1. Whereas the *ppx, abx, bx, pbx, bxd,* and *Ubx* sites all map in the left half of the BX-C region, the *iab*-series genes (together with *Hab* etc) map in the right half. Recent data (Morata, quoted in North 1984) suggests that the right half of the BX-C may be subdivided into two functional domains designated *abdA* (governing development of posterior A1 to A4) and *AbdB* (governing development of A5 to A8). Each of these domains includes several subordinate *iab* loci, just as the *ppx, abx, bx, pbx* and *bxd* loci form parts of the *Ubx* domain governing development of posterior T2 to anterior A1.

E. B. Lewis (1978) has accounted for these various mutant phenotypes by an elegantly simple model in which more and more of the bithorax genes are activated in successively more posterior segments. Thus none of the bithorax genes is active in the anterior compartment of T2 – the ground state. The Ubx^+ group of functions is active in T3, with bx^+ specifically required in anterior T3, and pbx^+ in posterior T3; bxd^+ activity is required in anterior A1 in addition to the above. This proviso follows from the observation that in *bxd/bxd* mutants, A1 is transformed towards T3 and not to the T2-like ground-state. Likewise, $iab2^+$ activity is required in A2 in addition to the BX-C genes already active in A1. Similar arguments can be advanced for the wild-type products of the other *iab*-series genes in more posterior abdominal segments, though these genes are as yet less fully characterised (see Lewis, 1978).

According to this model, the dominant gain-of-function mutants described earlier in this section could result from the inappropriate expression of certain BX-C loci in segments or compartments where they would normally remain inactive. Thus the *Contrabithorax* (*Cbx*) mutation might damage a regulatory site adjacent to the wild-type Ubx^+ locus, giving Ubx^+ expression in T2 (where it is inappropriate) as well as in T3 and abdominal segments, thereby converting T2 towards T3. Other mutants of this type (such as *Hab*) can be interpreted in a similar manner (Lewis, 1978).

The Lewis model implies that expression of all the BX-C genes must be repressed in T2, or at least in its anterior compartment. Derepression of the ppx^+ function may distinguish posterior T2 from posterior T1 (see above). The abx^+ and bx^+ elements of the *Ubx* domain would escape from repression in anterior T3, and likewise the pbx^+ site in posterior T3, while other BX-C genes remain inactive (fig. 78). In anterior A1, the bxd^+ gene is also derepressed, while the iab^+ genes are not. Finally, in A8, all the BX-C genes are active (fig. 78). Lewis (1978) has suggested that a single repressor substance distributed in a gradient from T2 (high) to A8 (low) could regulate the expression of all the BX-C genes by providing them with 'positional information', so that an appropriate pattern of activity is established in each segment or compartment. This would be a sufficient explanation provided that there is also a gradation in the strength of interaction between the repressor molecule and the regulatory sites of individual genes in the bithorax complex. In other words, different threshold levels of repressor would be required to inactivate each of these genes. A high threshold requirement for the inhibition of ppx^+, abx^+ and bx^+ expression would mean that a small drop in repressor level (from anterior T2 to posterior T2 and anterior T3) could derepress these genes, while the other BX-C genes would still remain inactive. The pbx^+ site would likewise escape from repressor influence in the posterior compart-

Elements of the BX–C	Genes ($\bullet$ = on, $\circ$ = off)	Anterior ⟶ Segment boundaries according to in-register model ⟵ Posterior									
		T2	T3	A1	A2	A3	A4	A5	A6	A7	A8
1	*Ubx* domain { $abx,^+ bx^+$ (ant. T3) pbx^+ (post. T3)	$\circ$	$\bullet$	$\bullet$	$\bullet$	$\bullet$	$\bullet$	$\bullet$	$\bullet$	$\bullet$	$\bullet$
2	bxd^+	$\circ$	$\circ$	$\bullet$	$\bullet$	$\bullet$	$\bullet$	$\bullet$	$\bullet$	$\bullet$	$\bullet$
3	*iab* 2^+	$\circ$	$\circ$	$\circ$	$\bullet$	$\bullet$	$\bullet$	$\bullet$	$\bullet$	$\bullet$	$\bullet$
4	[*iab* 3^+] *	$\circ$	$\circ$	$\circ$	$\circ$	$\bullet$	$\bullet$	$\bullet$	$\bullet$	$\bullet$	$\bullet$
5	[*iab* 4^+] *	$\circ$	$\circ$	$\circ$	$\circ$	$\circ$	$\bullet$	$\bullet$	$\bullet$	$\bullet$	$\bullet$
6	*iab* 5^+	$\circ$	$\circ$	$\circ$	$\circ$	$\circ$	$\circ$	$\bullet$	$\bullet$	$\bullet$	$\bullet$
7	[*iab* 6^+] *	$\circ$	$\circ$	$\circ$	$\circ$	$\circ$	$\circ$	$\circ$	$\bullet$	$\bullet$	$\bullet$
8	[*iab* 7^+] *	$\circ$	$\circ$	$\circ$	$\circ$	$\circ$	$\circ$	$\circ$	$\circ$	$\bullet$	$\bullet$
9	*iab* 8^+	$\circ$	$\circ$	$\circ$	$\circ$	$\circ$	$\circ$	$\circ$	$\circ$	$\circ$	$\bullet$

|A|P|A|P|A|P|A|P|A|P| (etc ?)
|T1|T2|T3|A1|A2|

Segment boundaries according to out-of-register model†

Notes *Genes in square brackets remain uncharacterised or poorly-characterised.
†In the out-of-register model, pbx^+ (required in posterior but not anterior T3) would be grouped with bxd^+ as element 2 of the BX–C. Similarly, ppx^+ (not included above) would be grouped with abx^+ and bx^+ as element 1.

Fig. 78 Proposed model for BX-C gene action (after Lewis, 1978).

ment of segment T3. Similarly, a lower threshold requirement for bxd^+ inhibition would mean that this gene stays repressed in T3 as well as T2, but becomes active in A1 and the other abdominal segments (fig. 78). This basic model can be varied slightly to account for segment-specific (in-register), out-of-register, or compartment-specific domains for the activity of individual BX-C genes.

Several genes have been identified which are involved in the regulation of the BX-C and other homoeotic functions, though none of them seems likely to encode the proposed BX-C repressor (gradient). When the wild-type product of the *Polycomb* (*Pc*) gene is lacking (in *Pc*⁻ mutants), then most of the body segments are partially transformed towards A8. This is the phenotype predicted if the proposed repressor were deficient or if the positional information it provides could not be interpreted appropriately; in either case, this would result in the indiscriminate activity of all the BX-C genes in all segments of the body. The Pc^+ gene product is required continuously throughout development to ensure a correct pattern of selector gene activity in the appropriate segments. This can be shown using temperature-sensitive *Pc*⁻ mutants, in which the wild-type gene product is unstable or otherwise deficient at the restrictive temperature. Abnormal development ensues in such mutants at whatever stage they are exposed to restrictive conditions. Another gene product (from the esc^+ gene) is required for correct expression of the BX-C and other homoeotic genes (see §7.6.4 below), but this is only required during early embryonic development. Thus in temperature-sensitive esc^- mutants there is a limited period during embryogenesis when restrictive conditions can disrupt development, implying that the wild-type gene product is only required at this early stage. Later in development the esc^+ product is no longer required, and restrictive conditions have no effect.

At least two further gene products are implicated in the regulation of BX-C and other selector genes, namely those defined by mutations in the *super sex combs* (*sxc*; Ingham, 1984) and *trithorax* loci (*trx*; Ingham, 1983). The wild-type sxc^+ gene product seems to act in a similar way to the esc^+ and Pc^+ products during early embryogenesis. Because the sxc^+ gene product is synthesised and accumulated in the oocyte, the effects of a deficiency in this product at early stages can only be studied in the absence of presynthesised maternal stores, i.e. in embryos derived from *sxc/sxc* mutant oocytes. Under such conditions, most of the embryonic segments become transformed towards A8. Later in development, the absence of sxc^+ gene product results in much less radical transformations, suggesting a function distinct from those of the esc^+ and Pc^+ gene products.

The wild-type trx^+ gene product, by contrast, seems to be required as an *activator* of the BX-C and other selector genes during later development. Thus *trx/trx* mutants show partial transformations of many segments towards the T2 condition (a converse pattern to the effect of Pc^-, sxc^- or esc^- mutations). To complicate the situation still further, homozygous combinations of esc^- and trx^- mutations restore an apparently normal pattern of segments (Ingham, 1983), implying that genes in the bithorax complex can be expressed differentially when both esc^+ and trx^+ gene products are absent. This would be difficult to explain if either the trx^+ product (activator) or esc^+ product (repressor) formed a gradient which directly specified selector gene activities. Rather, these two products may mediate in the interpretation of positional information generated by a separate (as yet uncharacterised) system. This would allow the effects of trx^- and esc^- deficiencies to cancel each other out, provided that other elements in the interpretation system can function to some extent in the absence of both esc^+ and trx^+ products.

Although the Lewis model is probably an oversimplification in some respects, it does make sense in evolutionary terms. While Dipterans have two wings and two halteres, most other insect groups have four wings;

thus some means for repressing the second pair of wings and substituting halteres on T3 must have evolved in the Diptera. Similarly the evolution of insects from primitive millipede-like arthropods would necessitate a mechanism for repressing the production of legs etc. by those segments which form the insect abdomen. These inhibitory functions are provided by the genes of the BX-C in Lewis's model.

Indications that there are further complications hidden within the bithorax complex have emerged from recent studies in which the left half of this complex (the *Ubx* domain) has been cloned as a series of overlapping DNA segments spanning 195 kbp (Bender *et al.*, 1983). Many of the known mutations affecting BX-C functions have also been mapped within this stretch of DNA (fig. 79); note that the whole left half of the BX-C locus is located in a small doublet band at region 89E of chromosome 3.

Most of the BX-C gene-cluster seems to be organised linearly from left to right in the antero-posterior order of gene expression, although *pbx* (required in posterior T3) is located to the right rather than to the left of *bxd* (required in anterior A1; see fig. 79). A relatively short RNA species is derived from the *bxd* transcription unit, which nevertheless spans some 20 kbp of DNA (-23 to -3 kbp on the BX-C map in fig. 79), as defined by the sites of insertions and chromosomal breakpoints causing *bxd*$^-$ mutations (see Bender *et al.*, 1983; North, 1983). The *Ubx* locus is even more extraordinary; apparently a 3.0 kb *Ubx* RNA is derived from two major exons which lie some 70 kbp apart in the BX-C map (fig. 79; North, 1983). Several puzzling features of *Ubx*$^-$ mutant phenotypes can be explained if *Ubx*$^-$ mutations inactivate the entire left half of the BX-C, including the *ppx*$^+$, *abx*$^+$, *bx*$^+$, *pbx*$^+$ and *bxd*$^+$ functions; thus *Ubx*$^-$ mutants show complete conversion of T3 and partial transformation of A1 towards a T2-like state. The *Ubx* map indeed implies that a huge primary transcript is produced from the left-hand end of this complex (*Ubx* domain) spanning both of the *Ubx* exon sites shown in fig. 79. There are probably alternative splicing pathways for this giant transcript, since several different-sized RNA species have been detected which share the same 5'-exon as the 3 kb *Ubx* RNA, but have different 3'-exons. These might include, for example, the site defined by *bx*$^-$ mutations, which map between the two *Ubx* sites (i.e. within the same transcription unit), and which cause a partial *Ubx*$^-$ phenotype (only the anterior T3 compartment being transformed towards T2). The *abx* site mentioned earlier also lies within the *Ubx* transcription unit, mapping to the left of the *bx* site.

Recently, the pattern of *Ubx* expression in different regions of the embryo and larva has been analysed by *in situ* hybridisation (Akam,

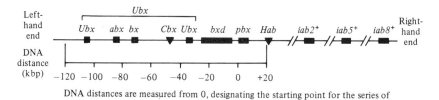

DNA distances are measured from 0, designating the starting point for the series of ovelapping clones

Dominant gain-of-function mutants are distinguished by the symbol ▼

Fig. 79 Genetic map of the bithorax complex in region 89E of chromosome 3 (after Bender *et al.*, 1983, and North, 1983).

1983). *Ubx* transcripts are detectable in both larval tissues and imaginal discs, particularly in segments T3 and A1 but not in T2 (just as predicted by the Lewis model). *Ubx* expression is prominent in both mesoderm and neural tissue as well as epidermis, consistent with the evidence cited earlier (end of §7.5) that BX-C gene products autonomously influence the segmental patterns of both muscles and nervous system in T3 and abdominal segments.

Bender *et al.* (1983) have shown that most of the X-ray-induced mutations in the bithorax complex result from significant deletions of DNA, whereas many of the spontaneous mutations affecting this complex are caused by insertions of a 7.3 kbp transposable element (§1.10) called 'gypsy', which is flanked by 500 bp direct repeats. These authors also suggest that the phenotype of the *Cbx¹* mutant (transforming posterior T2 into posterior T3) is the exact inverse of the *pbx¹* mutant. In fact, both of these mutations originally arose together after X-ray exposure, and were later separated by recombination. These facts can be neatly explained in molecular terms; the *pbx¹* mutation involves a 17 kbp deletion (−3 to +14 kbp on the map; fig. 79), whereas in the complementary *Cbx¹* mutation this same 17 kbp segment of DNA is found inserted in the opposite orientation at a different site mapping to −44 kbp. This brings the *pbx* DNA segment into the middle of the *Ubx* transcription unit, possibly activating it inappropriately (other evidence also suggests that *pbx* may be a regulatory rather than structural gene). In view of this new molecular information, it would now seem inappropriate to interpret some bithorax functions in terms of independent 'genes', since alternative splicing pathways may provide more than one functional product from a single long transcription unit (e.g. *Ubx* and *bx*?). The regulation of this left-hand end of the bithorax locus should prove intriguing, but probably complex indeed!

7.6.4 Other homoeotic loci and their interactions

Two different homoeotic mutations normally produce additive effects on the adult phenotype, or on the segmental cuticle patterns of the larvae. Examples have been met with in the previous section, for example the complete transformation of T3 into T2 in *abx⁻bx⁻pbx⁻* homozygotes, or the transformation of T3 and most abdominal segments towards a 'T2-like' ground state in mutant larvae lacking the entire bithorax complex (BX-C⁻). Similarly, the effects of *engrailed* and bithorax mutant alleles are additive; a particularly clear example of this was described by Garcia-Bellido *et al.* (1979) for a homozygous combination of the *engrailed* and *postbithorax* mutations (*pbx en/pbx en*). Here the posterior compartment of the adult wing on T2 is transformed into a mirror-image of the anterior compartment (phenotype of *en/en*; see §7.6.2 above), and the posterior part of the haltere on T3 is transformed towards wing (phenotype of *pbx/pbx*; see §7.6.3 above). But in addition, the venation pattern and presence of marginal bristles on the wing-like portion of the T3 haltere suggest that this posterior region is in fact transformed into a mirror-image anterior wing compartment, resulting from the combined effects of both mutations together.

As discussed in the previous section, *Polycomb* mutants in which the wild-type *Pc⁺* gene product is lacking show transformation of all the larval segments towards A8, suggesting that the BX-C genes are active indiscriminately throughout the embryo (Lewis, 1978). A similar pattern is found in null mutants lacking the wild-type product of the *extra sex combs* (*esc⁺*) gene. Nevertheless, these two genes (*Pc⁺* and *esc⁺*) act independently; for one thing *esc⁺* activity is required only during early embryogenesis, whereas *Pc⁺* activity is required throughout development (see §7.6.3 above; Struhl, 1981b; Struhl & Brower, 1982; Struhl, 1983). The effects of combining *esc⁻* and BX-C⁻ mutations are shown schematically in fig. 80 (after Lawrence, 1981a). Note that in BX-C⁻ embryos retaining the *esc⁺* function, the characteristic transformation of T3 and abdominal segments towards the ground state is incomplete at the posterior end, in that A8 shows a T′ character intermediate between prothorax (T1) and mesothorax (T2), while a normally latent ninth abdominal segment (A9) is expressed but not transformed towards thorax. In BX-C⁻ *esc⁻* double mutant embryos, the intermediate T′ character is apparent in all of the larval segments (fig. 80).

It will be noted that the *esc⁻* mutation affects both the head (H) and prothoracic (T1) segments, as well as the pattern of T2, T3 and abdominal segments specified by the bithorax complex. Thus all body segments are

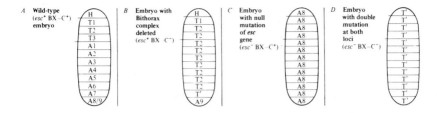

Fig. 80 Interaction of the *esc* and BX-C loci in *Drosophila* development (after Lawrence, 1981a).

transformed towards A8 in *esc⁻* BX-C⁺ embryos (implying that the BX-C genes are indiscriminately active even in T1 and H), while in *esc⁻* BX-C⁻ embryos the T′ pattern intermediate between T1 and T2 is likewise found in all segments. These results indicate that the wild-type *esc⁺* gene product is required to establish an appropriate pattern of gene activity not only for the BX-C genes but also for other selector genes, some of which govern anterior (T1 and head) development.

These other homoeotic genes include the *Antennapedia* (*Antp*) locus mentioned earlier (§7.6.1) and also the *Sex combs reduced* (*Scr*) gene. Both form part of another linked gene-complex designated ANT-C (Kaufman *et al.*, reported in North, 1983; Scott *et al.*, 1983), which also includes a non-homoeotic gene (*ftz*) affecting segment number, and at least one further homoeotic gene affecting head development (*proboscipedia*). The *Antp* transcription unit is at least 100 kb long (Scott *et al.*, 1983), and apparently this huge transcript can be processed via alternative splicing pathways to yield two developmentally regulated RNA species of different sizes (Scott *et al.*, 1983; Garber *et al.*, 1983). *In situ* hybridisation has been used to analyse the distribution of *Antp* transcripts in different regions of the embryo and larvae; most hybridisation is found in the larval nervous system and imaginal discs of the three thoracic segments (see Levine *et al.*, 1983). The roles of the *Antp⁺* and *Scr⁺* loci are at present less well-defined than those of the bithorax complex, and no complete description of the pattern of anterior development has yet emerged. Several further but as yet undefined head-determining genes would seem to be required, with homoeotic mutants such as *ophthalmoptera* and *proboscipedia* offering possible candidates.

Scr⁻ mutants transform the prothorax (T1) partially towards mesothorax (T2), but also transform the labial head segment towards the maxillary state. Thus *Scr⁺* activity (wild-type gene product) is probably required for normal development of the T1 and labial segments. As shown in fig. 80, all of the body segments in *esc⁻* BX-C⁻ homozygous embryos are transformed to a T′ phenotype intermediate between T1 and T2. This is consistent with the Scr^+ gene being expressed indiscriminately throughout the body rather than active only in theT1 and labial segments. Again this implies that appropriate expression of the Scr^+ gene is dependent on the esc^+ gene product during early development.

The distribution of $Antp^+$ activity in normal embryos is less easily defined; there are at least two possible interpretations, neither of which is completely satisfactory. What follows is a simplified version of that propounded by Struhl (1983). As mentioned earlier, null $Antp^-$ mutants produce antennae in place of second legs (Struhl, 1981a), whereas dominant mutations at this locus cause the reverse transformation of antennae into second legs. This suggests that the wild-type $Antp^+$ gene product is required in the mesothorax (T2) in order to prevent the inappropriate expression of antennal characteristics. If so, the $Antp^+$ gene must normally remain silent in the antennal disc, since antennal development proceeds only in the absence of the $Antp^+$ gene product. Hence the dominant $Antp$ mutants could result from inappropriate expression of the $Antp^+$ gene product in the antennal disc, giving a T2-like phenotype. There is also evidence that the $Antp^+$ gene product is required additionally in T1 and T3. Thus the first and third legs also show antennal characteristics in $Antp^-$ mutants which lack the Ubx^+ and Scr^+ gene products (Struhl, 1982). Notably, Ubx^- mutations convert T3 towards T2, while Scr^- mutations convert T1 towards T2. Hence in a situation where all three thoracic segments resemble T2, a lack of $Antp^+$ gene product becomes manifested in the conversion of all six legs towards antennae. This is entirely consistent with the idea mentioned earlier (§7.4), that the combinatorial activities of several selector (homoeotic) genes specify the detailed 'address' or determined state of particular groups of cells. The $Antp^+$ gene product may also be expressed in the first seven abdominal segments but not in the eighth, as suggested by the fact that in esc^- BX-C$^-$ embryos retaining the $Antp^+$ function, the phenotype of A8 is influenced by inappropriate expression of the $Antp^+$ gene, a feature not seen in esc^- BX-C$^-$ $Antp^-$ mutants (Struhl, 1983).

Thus the appropriate pattern of $Antp^+$ expression, like that of the Scr^+ and BX-C genes, may be established in the early embryo via the wild-type product of the esc^+ gene. Fig. 81 shows the model proposed by Struhl (1983) on the basis of the above data and accompanying interpretations. Whereas the esc^+ gene product is required only initially to establish the pattern of homoeotic (selector) gene expression in appropriate segments, the Pc^+ gene product is required continuously in order to maintain this pattern. The sxc^+ and trx^+ gene products described earlier (§7.6.3 above) are not included in this scheme, but should be classed along with the Pc^+ and esc^+ products as being expressed in all segments. It remains to be seen how the Pc^+, esc^+, sxc^+ (all 'repressors') and trx^+ ('activator') products interact with selector genes expressed in compartmentally restricted patterns. At least one additional head-determining homoeotic gene is required by this model (designated HHG in fig. 81), although the evidence for its existence is only indirect as yet. This scheme extends the Lewis model (fig. 78) to account for prothoracic and some aspects of head development. But equally, we are still far from a complete picture of how the different *Drosophila* segments and compartments are specified by homoeotic (selector) gene activities; particularly this is true for the head segments.

Recently, a common sequence (designated H or 'homoeo box') has been identified in several homoeotic genes (McGinnis *et al.*, 1984a; Akam, 1984). There are at least eight copies of this homoeo box sequence in the *Drosophila* genome, one of which lies within the *Antp* transcription unit and another within the *Ubx* transcription unit. Two further copies occur in the BX-C, one each in the *abdA* and *AbdB* domains (see North, 1984). Several more occur within the ANT-C, including one in the *ftz* gene (Laughon & Scott, 1984), as well as one in the *engrailed* gene which lies outside the BX-C and ANT-C regions (Kornberg & Gehring, quoted in North, 1984). These common sequences may encode a highly conserved protein segment essential to the function of selector gene products. Despite this rapid progress, it still remains (i) to identify the ultimate gene products encoded by homoeotic genes, (ii) to establish how these products work at a molecular level, (iii) to elucidate the regulation of these genes (e.g. the roles of the Pc^+, esc^+, sxc^+ and trx^+ gene products), and (iv) to define the system of positional information to which homoeotic genes respond with compartment-specific patterns of activity.

A more distant prospect is the extension of this selector-gene model to other animals, including vertebrates which also show some evidence of an underlying segmental organisation (e.g. in the vertebrae and spinal cord, and in the embryonic somite blocks). However, progress will inevitably be slow until we can characterise homoeotic mutants defining the selector genes. Recent work has identified a series of mutations at the *lin-12* locus in *Caenorhabditis elegans* which show homoeotic effects on binary determination decisions in development (Greenwald *et al.*, 1983). In vertebrates, the effects of homoeotic mutations might be less obvious, since segmental features are much obscured as compared with insects. Very recently, close relatives of the common H sequence or "homoeo box" (shared by several *Drosophila* homoeotic genes; see above) have been

	Segment	Controlling genes		'Anterior' genes			Elements of Bithorax complex								
		esc	*Pc*	*Antp*	*HHG**	*Scr*	1	2	3	4	5	6	7	8	9
Head segments	Labral	−	−	−	−	−	−	−	−	−	−	−	−	−	−
	Antennal	●	●	○	●	○	○	○	○	○	○	○	○	○	○
	Premandibular	−	−	−	−	−	−	−	−	−	−	−	−	−	−
	Mandibular	−	−	−	−	−	−	−	−	−	−	−	−	−	−
	Maxillary	●	●	○	●	○	○	○	○	○	○	○	○	○	○
	Labial	●	●	○	●	●	○	○	○	○	○	○	○	○	○
Thoracic segments	Prothorax T1	●	●	●	○	●	○	○	○	○	○	○	○	○	○
	Mesothorax T2	●	●	●	○	○	⊕§	○	○	○	○	○	○	○	○
	Metathorax T3	●	●	●	○	○	●	○	○	○	○	○	○	○	○
Abdominal segments	A1	●	●	●	○	○	●	●	○	○	○	○	○	○	○
	A2	●	●	●	○	○	●	●	●	○	○	○	○	○	○
	A3	●	●	●	○	○	●	●	●	●	○	○	○	○	○
	A4	●	●	●	○	○	●	●	●	●	●	○	○	○	○
	A5	●	●	●	○	○	●	●	●	●	●	●	○	○	○
	A6	●	●	●	○	○	●	●	●	●	●	●	●	○	○
	A7	●	●	●	○	○	●	●	●	●	●	●	●	●	○
	A8	●	●	○	⊘†	○	●	●	●	●	●	●	●	●	●
	A9	●	●	−	−	−	−	−	−	−	−	−	−	−	−

Gene product required only during early embryogenesis   Gene product required continuously throughout development

Expressed in all segments   Expressed in a segment-specific pattern

● Gene active   ○ Gene inactive   − Pattern of gene activity not defined

Fig. 81 Establishment and maintenance of segment-specific patterns of homoeotic gene expression by the esc^+ and Pc^+ gene products (modified from Struhl, 1983).
* *HHG* denotes a hypothetical head-determining gene for which there is as yet only indirect evidence.
† One or more anterior homoeotic genes may normally be active in A8, since deletion of the bithorax complex transforms A8 into T′ (intermediate T1/T2) rather than T2.
§ Some element of the *Ubx* domain is expressed in posterior T2. See text and Fig. 78 for explanation of the out-of-register model for BX-C gene action.

identified in the genomes of *Xenopus* (Carrasco *et al.*, 1984) and perhaps of chicken and man, as well as in segmented creatures such as earthworm and beetle (McGinnis *et al.*, 1984b; Struhl, 1984b). If indeed these conserved 'homoeo box' (H) sequences imply some sort of homoeotic function for genes containing them, then it should be possible to short-circuit their analysis in vertebrates. Briefly, the identified H sequences would be used as starting points for cloning the DNA surrounding them. The transcription of these sequences during embryonic development could then be examined by Northern blotting analysis (3:2) with the cloned probes, and by *in situ* hybridisation to determine the spatial distribution of the transcripts, e.g. in the embryonic somites and neural tube. This approach has already shown that a *Xenopus* gene containing a homoeo-box sequence is only expressed during early embryonic development, producing three transcripts in a defined temporal order starting late in gastrulation (Carrasco *et al.*, 1984). The possibility of a universal genetic mechanism governing body plan in diverse animals (Struhl, 1984b) is indeed an exciting one. Such studies may also provide clues to the mechanism of action of homoeotic (selector) genes. For instance, the conserved 'homoeo box' sequence encodes a highly-basic protein domain which shows significant amino-acid homologies to the DNA-binding regions of certain procaryotic DNA-binding proteins, and to parts of the two *MATα* genes involved in switches of mating-type in yeast (see Laughon & Scott, 1984; Shepherd *et al.*, 1984). Since the product specified by a selector gene must presumably affect the expression of many other genes within its domain of action (i.e. in a particular segment or compartment), such a DNA-binding function might prove important for selector-gene action in development. Recently, it has proved possible to identify proteins encoded by the *Ubx* locus, using antibodies against these proteins as expressed in bacterial cells containing cloned *Ubx* cDNA sequences. These *Ubx* proteins are largely localised in cell nuclei during *Drosophila* embryonic and larval development, again suggesting a DNA-binding function (White & Hogness, quoted in North, 1984). No doubt many further details of homoeotic gene action and control will emerge in the near future, both in *Drosophila* and other animal systems.

References

Adesnik, M. & Darnell Jr., J. E. (1972). Biogenesis and characterisation of histone messenger RNA in Hela cells. *J. molec. Biol.* **67**, 397–406.

Agata, K., Yasuda, K. & Okada, T. S. (1983). Gene coding for a lens-specific protein, δ crystallin, is transcribed in nonlens tissues of chicken embryos. *Devl Biol.* **100**, 222–6.

Akam, M. (1983). The location of *Ultrabithorax* transcripts in *Drosophila* tissue sections. *EMBO J.* **2**, 2075–84.

Akam, M. (1984). A common segment in genes for segments of *Drosophila*. *Nature, Lond.* **308**, 402–3.

Alberts, B., Bray, D., Lewis, J., Raff, M., Roberts, K. & Watson, J. D. (1983). *Molecular Biology of the Cell*. Garland Publishing Inc., New York & London.

Alwine, J. C., Kemp, D. J. & Stark, G. R. (1977). Method for detection of specific RNAs in agarose gels by transfer to diazobenzyloxymethyl-paper and hybridisation with DNA probes. *Proc. natn. Acad. Sci. USA*, **74**, 5350–4.

Angelier, N. & Lacroix, J. C. (1975). Complexes de transcription d'origines nucleolaire et chromosomique d'ovocytes de *Pleurodeles waltlii* et *P. poireti* (Amphibiens, Urodeles). *Chromosoma*, **51**, 323–35.

Arndt-Jovin, D. J., Robert-Nicoud, M., Zarling, D. A.. Greider, C., Weiner, E. & Jovin, T. M. (1983). Left-handed Z-DNA in bands of acid-fixed polytene chromosomes. *Proc. natn. Acad. Sci. USA*, **80**, 4344–8.

Arst, H. N. & Macdonald, D. W. (1975). A gene cluster in *Aspergillus nidulans* with an internally located *cis*-acting regulatory region. *Nature, Lond.* **254**, 26–31.

Ashburner, M. (1980). Chromosomal action of ecdysone. *Nature, Lond.* **285**, 435–6.

Ashburner, M., Chihara, C., Meltzer, P. & Richards, G. (1973). Temporal control of puffing activity in polytene chromosomes. *Cold Spring Harb. Symp. quant. Biol* **38**, 655–62.

Avvedimento, V. E., Vogeli, G., Yamada, Y., Maizel Jr., J. V., Pastan, I. & de Crombrugghe, B. (1980). Correlation between splicing sites within an intron and their sequence complementarity with U1RNA. *Cell*, **21**, 689–96.

Axel, R., Feigelson, P. & Schutz, G. (1976). Analysis of the complexity and diversity of mRNA from chicken oviduct and liver. *Cell*, **7**, 247–54.

Babich, A., Nevins, J. R. & Darnell, J. E. (1980). Early capping of transcripts from the adenovirus major late transcription unit. *Nature, Lond.* **287**, 246–8.

250

Balhorn, R., Jackson, D., Granner, D. & Chalkley, R. (1975). Phosphorylation of the lysine-rich histones throughout the cell cycle. *Biochemistry*, **14**, 2504–11.

Banerji, J., Olson, L. & Schaffner, W. (1983). A lymphocyte-specific cellular enhancer is located downstream of the joining region in immunoglobulin heavy chain genes. *Cell*, **33**, 729–40.

Baralle, F. E., Shoulders, C. C. & Proudfoot, N. J. (1980). The primary structure of the human ε globin gene. *Cell*, **21**, 621–6.

Beach, L. R. & Palmiter, R. D. (1981). Amplification of the metallothionein I gene in cadmium-resistant mouse cells. *Proc. natn. Acad. Sci. USA*, **78**, 2110–14.

Beebee, T. J. C. & Butterworth, P. H. W. (1974). Template specificities of *X. laevis* RNA polymerases: selective transcription of ribosomal cistrons by RNA polymerase A. *Eur. J. Biochem.* **45**, 395–406.

Beerman, W. (1964). Control of differentiation at the chromosomal level. *J. exp. Zool.* **157**, 49–62.

Benbow, R. M., Pestell, R. Q. W. & Ford, C. C. (1975). Appearance of DNA polymerase activities during early development of *Xenopus laevis*. *Devl Biol.* **43**, 159–74.

Bender, W., Akam, M., Karch, F., Beachy, P. A., Peifer, M., Spierer, P., Lewis, E. B. & Hogness, D. S. (1983). Molecular genetics of the bithorax complex in *Drosophila melanogaster*. *Science*, **221**, 23–9.

Bentinnen, L. C. & Comb, D. G. (1971). Early and late histones during sea urchin development. *J. molec. Biol.* **57**, 355–8.

Berger, S. L. & Cooper, H. L. (1975). Very short-lived and stable mRNAs from resting human lymphocytes. *Proc. natn. Acad. Sci. USA*, **72**, 3873–7.

Bernard, O., Hozumi, N. & Tonegawa, S. (1978). Sequences of mouse immunoglobulin light chain genes before and after somatic changes. *Cell*, **15**, 1133–44.

Bhorjee, J. (1981). Differential phosphorylation of nuclear nonhistone high mobility group proteins HMG 14 and 17 during the cell cycle. *Proc. natn. Acad. Sci. USA*, **78**, 6944–8.

Bier, K. (1963). Syntheses, interzellulärer transport, und abbua von ribonukleinsäure in ovar der stubenfliege *Musca domestica*. *J. Cell Biol.* **16**, 436–40.

Bird, A. P. (1984). DNA methylation – how important in gene control? *Nature, Lond.* **307**, 503–4.

Bird, A., Taggart, M. & Macleod, D. (1981). Loss of rDNA methylation accompanies the onset of ribosomal gene activity in early development of *X. laevis*. *Cell*, **26**, 381–90.

Birnstiel, M., Telford, J., Weinberg, E. & Stafford, D. (1974). Isolation and some properties of the genes coding for histone proteins. *Proc. natn. Acad. Sci. USA*, **71**, 2900–4.

Bishop, J. O., Pemberton, R. & Baglioni, C. (1972). Reiteration frequency of haemoglobin genes in the duck. *Nature New Biol.* **235**, 231–4.

Blake, C. C. F. (1978). Do genes-in-pieces imply proteins-in-pieces? *Nature, Lond.* **273**, 267–8.

Blake, C. (1983). Exons – present from the beginning? *Nature, Lond.* **306**, 535–7.

Bogenhagen, D. F., Sakonju, S. & Brown, D. D. (1980). A control region in the centre of the 5S gene directs specific initiation of transcription: II, the 3′ border of the region. *Cell*, **19**, 27–35.

Bonner, J. J. & Pardue, M.-L. (1977a). Polytene chromosome puffing and *in situ* hybridisation measure different aspects of RNA metabolism. *Cell*, **12**, 227–34.

Bonner, J. J. & Pardue, M.-L. (1977b). Ecdysone-stimulated RNA synthesis in salivary glands of *Drosophila melanogaster:* assay by *in situ* hybridisation. *Cell*, **12**, 219–25.

Boseley, P., Moss, T., Machler, M., Portmann, R. & Birnstiel, M. L. (1979). Sequence organisation of the spacer DNA in a ribosomal gene unit of *Xenopus laevis. Cell*, **17**, 19–31.

Boveri, T. (1899). Die entwicklung von *Ascaris megalocephala* mit besonderer rücksicht auf die kernverhältnisse, p. 383. F. C. von Kupffer, Jena.

Boycott, A. E., Diver, C., Garstrang, S. L. & Turner, F. M. (1930). The inheritance of sinistrality in *Limnaea peregra* (Mollusca, Pulmonata). *Phil. Trans. R. Soc. Ser. B*, **219**, 51–131.

Bradbury, E. M., Inglis, R. J. & Matthews, H. R. (1974a). Control of cell division by very lysine-rich histone (F1) phosphorylation. *Nature, Lond.* **247**, 257–61.

Bradbury, E. M., Inglis, J. R., Matthews, H. R. & Langan, T. (1974b). Molecular basis of control of mitotic cell division in eucaryotes. *Nature, Lond.* **249**, 535–5.

Bradbury, E. M., Maclean, N. & Matthews, R. (1981). *DNA, Chromatin and Chromosomes*. Blackwell Scientific Publications, Oxford.

Brandhorst, B. P. & McConkey, E. H. (1974). Stability of nuclear RNA in mammalian cells. *J. molec. Biol.* **85**, 451–63.

Brandis, J. W. & Raff, R. A. (1978). Translation of oogenetic mRNA in sea urchin eggs and early embryos; demonstration of a change in translational efficiency following fertilization. *Devl Biol.* **67**, 99–113.

Breathnach, R., Benoist, C., O'Hare, K., Gannon, F. & Chambon, P. (1978). Ovalbumin gene: evidence for a leader sequence in mRNA and DNA sequences at the exon-intron boundaries. *Proc. natn. Acad. Sci. USA*, **75**, 4853–7.

Brenner, S. (1974). The genetics of *Caenorhabditis elegans. Genetics*, **77**, 71–94.

Briggs, R., Green, E. V. & King, T. J. (1951). An investigation of the capacity of cleavage and differentiation in *Rana pipiens* eggs lacking 'functional' chromosomes. *J. exp. Zool.* **116**, 455-94.

Briggs, R. & King, T. J. (1957). Changes in the nuclei of differentiating endoderm cells as revealed by nuclear transplantation. *J. Morph.* **100**, 269–312.

Britten, R. J. & Davidson, E. H. (1969). Gene regulation for higher cells: a theory. *Science*, **165**, 349–56.

Britten, R. J. & Kohne, D. E. (1968). Nucleotide sequence repetition in DNA. *Science*, **161**, 529–40.

Brock, M. L. & Shapiro, D. J. (1983). Estrogen stabilises vitellogenin mRNA against cytoplasmic degradation. *Cell*, **34**, 207–14.

Brower, D. L., Smith, R. J. & Wilcox, M. (1982). Cell shapes on the surface of the *Drosophila* wing imaginal disc. *J. Embryol. exp. Morph.* **67**, 137–51.

Brown, D. D. & Gurdon, J. B. (1977). Cloned single repeating units of 5S DNA direct accurate transcription of 5S RNA when injected into *Xenopus* oocytes. *Proc. natn. Acad. Sci. USA*, **74**, 2064–8.

Brown, D. D., Wensink, P. C. & Jordan, E. (1972). A comparison of the ribosomal DNA of *X. laevis* and *X. mulleri:* the evolution of tandem genes *J. molec. Biol.* **63**, 57–73.

Bryant, P. J. (1971). Regeneration and duplication following operations *in situ* on the imaginal discs of *Drosophila melanogaster. Devl Biol.* **26**, 606–15.

Burch, J. B. E. & Weintraub, H. (1983). Temporal order of chromatin structural changes associated with activation of the major chicken vitellogenin gene. *Cell,* **33**, 65–76.

Burns, A. T. H., Deeley, R. G., Gordon, J. I., Udell, D. S., Mullinix, K. P. & Goldberger, R. F. (1978). Primary induction of vitellogenin mRNA in the rooster by 17-β-oestradiol. *Proc. natn. Acad. Sci. USA,* **75**, 1815–19.

Busby, S. J. & Reeder, R. H. (1983). Spacer sequences regulate transcription of ribosomal gene plasmids injected into *Xenopus* embryos. *Cell,* **34**, 989–96.

Busch, H., Reddy, R., Rothblum, L. & Choi, Y. C. (1982). snRNAs, snRNPs, and RNA processing. *A. Rev. Biochem.* **51**, 617–54.

Busslinger, M., Hurst, J. & Flavell, R. A. (1983). DNA methylation and the regulation of globin gene expression. *Cell,* **34**, 197–206.

Callan, H. G. (1963). The nature of lampbrush chromosomes. *Int. Rev. Cytol.* **15**, 1–34.

Capco, D. G. & Jeffery, W. R. (1981). Regional accumulation of vegetal pole poly(A)$^+$ RNA injected into fertilised *Xenopus* eggs. *Nature, Lond.* **294**, 255–7.

Capco, D. G. & Jeffery, W. R. (1982). Transient localisations of messenger RNA in *Xenopus laevis* oocytes. *Devl Biol.* **89**, 1–12.

Carrasco, A. E., McGinnis, W., Gehring, W. J. & De Robertis, E. M. (1984). Cloning of an *X. laevis* gene expressed during early embryogenesis coding for a peptide region homologous to *Drosophila* homoeotic genes. *Cell,* **37**, 409–14.

Carroll, D. & Brown, D. D. (1976). Repeating units of *Xenopus laevis* oocyte type 5S DNA are heterogeneous in length. *Cell,* **7**, 467–75.

Case, S. T. & Daneholt, B. (1977). Cellular and molecular aspects of genetic expression in *Chironomus* salivary glands. In *Biochemistry of Cell Differentiation II*, vol. 15. ed. J. Paul, pp. 45–77. University Park Press, Baltimore.

Catterall, J. F., O'Malley, B. W., Robertson, M. A., Staden, R., Tanaka, Y. & Brownlee, G. G. (1978). Nucleotide sequences at 12 intron-exon junctions in the chick ovalbumin gene. *Nature, Lond.* **275**, 510–13.

Catterall, J. F., Stein, J. P., Lai, E. C., Woo, S. L. C., Dugaiczyk, A., Mace, M. L., Means, A. R. & O'Malley, B. W. (1979). The chick ovomucoid gene contains at least six intervening sequences. *Nature, Lond.* **278**, 323–7.

Cech, T. R. (1983). RNA splicing: three themes with variations. *Cell,* **34**, 713–16.

Cech, T. & Hearst, J. E. (1975). An electron microscopic study of mouse foldback DNA. *Cell,* **5**, 429–46.

Cech, T., Potter, D. & Pardue, M.-L. (1977). Electron microscopy of DNA cross linked with trimethylpsoralen: A probe for chromatin structure. *Biochemistry,* **16**, 5313–20.

Chantrenne, H., Burny, A. & Marbaix, G. (1967). The search for mRNA of hemoglobin. *Prog. nucl. Acids Res. & mol. Biol.* **7**, 173–94.

Chapman, B. S. & Tobin, A. J. (1979). Distribution of developmentally-regulated haemoglobins in embryonic erythroid populations. *Devl Biol.* **69**, 375–87.

Chikaraishi, D. M., Deeb, S. S. & Sueoka, N. (1978). Sequence complexity of nuclear RNAs in adult rat tissues. *Cell*, **13**, 111–20.

Ciliberto, G., Raugei, G., Costanzo, F., Dente, L. & Cortese, R. (1983). Common and interchangeable elements in the promoters of genes transcribed by RNA polymerase III. *Cell*, **32**, 725–33.

Clayton, R. M., Thomson, I. & de Pomerai, D. I. (1979). Relationship between crystallin mRNA expression in retina cells and their capacity to re-differentiate into lens cells. *Nature, Lond.* **282**, 628–9.

Clement, A. C. (1962). Development of *Ilyanassa* following removal of the D macromere at successive cleavage stages. *J. exp. Zool.* **149**, 193–215.

Cleveland, D. W. (1983). The tubulins: from DNA to RNA to protein and back again. *Cell*, **34**, 330–2.

Cochet, M., Chang, A. C. Y. & Cohen, S. N. (1982). Characterisation of the structural gene and putative 5′ regulatory sequences for human pro-opiomelanocortin. *Nature, Lond.* **297**, 335–8.

Cochet, M., Gannon, F., Hen, R., Maroteaux, L., Perrin, F. & Chambon, P. (1979). Organisation and sequence studies of the 17-piece chicken conalbumin gene. *Nature, Lond.* **282**, 567–74.

Cohn, R. H. & Kedes, L. H. (1979). Nonallelic histone gene clusters of individual sea urchins (*L. pictus*): polarity and gene organisation. *Cell*, **18**, 843–53.

Compton, J. G., Schrader, W. T. & O'Malley, B. W. (1983). DNA sequence preference of the progesterone receptor. *Proc. natn. Acad. Sci. USA*, **80**, 16–20.

Conklin, E. G. (1905). The organisation and cell lineage of the ascidean egg. *J. Acad. nat. Sci. Philad.* **13**, 5–118.

Corden, J., Wasylyk, B., Buchwalder, A., Sassone-Cordi, P., Kedinger, C. & Chambon, P. (1980). Promoter sequences of eucaryotic protein-coding genes. *Science*, **209**, 1406–14.

Corneo, G., Ginelli, E. & Polli, E. (1970). Different satellite DNA of guinea pig and ox. *Biochemistry*, **9**, 1565–71.

Cory, S., Jackson, J. & Adam, J. M. (1980). Deletions in the constant region locus can account for switches in immunoglobulin heavy chain expression. *Nature, Lond.* **284**, 450–3.

Costantini, F. D., Scheller, R. H., Britten, R. J. & Davidson, E. H. (1978). Repetitive sequence transcripts in the mature sea urchin oocyte. *Cell*, **15**, 173–87.

Coveney, J. & Woodland, H. R. (1982). The DNase I sensitivity of *Xenopus laevis* genes transcribed by RNA polymerase III. *Nature, Lond.* **298**, 578–10.

Cozzarelli, N. R., Gerrard, S. P., Schlissel, M., Brown, D. D. & Bogenhagen, D. F. (1983). Purified RNA polymerase III accurately and efficiently terminates transcription of 5S RNA genes. *Cell*, **34**, 829–35.

Craik, C. S., Buchman, S. R. & Beychok, S. (1980). Characterisation of globin domains: heme binding to the central exon product. *Proc. natn. Acad. Sci. USA*, **77**, 1384–8.

Cramer, J. H., Sebastian, J., Rownd, R. H. & Halvorson, H. O. (1974). Transcription of *Saccharomyces cerevisiae* ribosomal DNA *in vivo* and *in vitro*. *Proc. natn. Acad. Sci. USA*, **71**, 2188–92.

Crick, F. H. C. & Lawrence, P. A. (1975). Compartments and polyclones in insect development. *Science*, **189**, 340–7.

Crowley, T. E., Bond, M. W. & Meyerowitz, E. M. (1983). The structural genes for three *Drosophila* glue proteins reside at a single polytene chromosome puff locus. *Mol. Cell. Biol.* **3**, 623–34.

Cudennec, C. A., Thiery, J.-P. & le Douarin, N. M. (1981). *In vitro* induction of adult erythropoiesis in early mouse yolk sac. *Proc. natn. Acad. Sci. USA*, **78**, 2412–16.

Curtis, P. J., Mantei, N., van den Berg, J. & Weissman, C. (1977). Presence of a putative 15S precursor to β globin mRNA but not to α globin mRNA in Friend cells. *Proc. natn. Acad. Sci. USA*, **74**, 3184–8.

Daneholt, B., Andersson, K. & Fagerlind, M. (1977). Large sized polysomes in *Chironomus tentans* salivary glands and their relation to Balbiani ring 75S RNA. *J. Cell Biol.* **73**, 149–60.

Daneholt, B., Case, S. T., Derksen, J., Lamb, M. M., Nelson, L. G. & Wieslander, L. (1979). The transcription unit in Balbiani ring 2 and its relation to the chromomeric subdivision of the polytene chromosome. In *Specific Eukaryotic Genes*, Alfred Bentzon Symp. **13**, 39–51.

Dan-Sohkawa, M. & Satoh, N. (1978). Studies on dwarf larvae from isolated blastomeres of the starfish *Asterina pectenifera. J. Embryol. exp. Morph.* **46**, 171–85.

Darnell Jr., J. E. (1982). Variety in the level of gene control in eukaryotic cells. *Nature, Lond.* **297**, 365–71.

Davidson, E. H. (1976). *Gene Activity in Early Development*, 2nd. ed. Academic Press, New York & London.

Davidson, E. H. & Britten, R. J. (1971). Note on the control of gene expression during development. *J. theor. Biol.* **32**, 123–30.

Davidson, E. H. & Britten, R. J. (1979). Regulation of gene expression: possible role of repetitive sequences. *Science*, **204**, 1052–9.

Davidson, E. H., Hough, B. R., Amenson, C. S. & Britten, R. J. (1973). General interspersion of repetitive with non-repetitive sequence elements in the DNA of *Xenopus. J. molec. Biol.* **77**, 1–24.

Davidson, E. H., Hough-Evans, B. R. & Britten, R. J. (1982). Molecular biology of the sea urchin embryo. *Science*, **217**, 17–26.

Davidson, E. H., Jacobs, H. T. & Britten, R. J. (1983). Very short repeats and coordinate induction of genes. *Nature, Lond.* **301**, 468–70.

Davidson, E. H., Klein, W. H. & Britten, R. J. (1977). Sequence organisation in animal DNA and a speculation on hnRNA as a coordinate regulatory transcript. *Devl Biol.* **55**, 69–84.

Davidson, E. H. & Posakony, J. W. (1982). Repetitive sequence transcripts in development. *Nature, Lond.* **297**, 633–5.

Dawid, I. B. & Wellauer, P. K. (1977). Ribosomal DNA and related sequences in *Drosophila melanogaster. Cold Spring Harb. Symp. quant. Biol.* **42**, 1185–94.

Dean, D. C., Knoll, B. J., Riser, M. E. & O'Malley, B. W. (1983). A 5′-flanking sequence essential for progesterone regulation of an ovalbumin fusion gene. *Nature, Lond.* **305**, 551–3.

Denison, R. A., van Arsdell, S. W., Bernstein, L. B. & Weiner, A. M. (1981). Abundant pseudogenes for small nuclear RNAs are dispersed in the human genome. *Proc. natn. Acad. Sci. USA*, **78**, 810–14.

Deppe, U., Schierenberg, E., Cole, T., Krieg, C., Schmitt, D., Yoder, B. & von Ehrenstein, G. (1978). Cell lineages of the embryo of the nematode *Caenorhabditis elegans. Proc. natn. Acad. Sci. USA*, **75**, 376–80.

Derksen, J., Wieslander, L., van der Ploeg, M. & Daneholt, B. (1980). Identification of the Balbiani ring 2 chromomere and determination of the content and compaction of its DNA. *Chromosoma*, **81**, 65–84.

Derman, E. & Darnell, Jr, J. E. (1974). Relationship of chain transcription to poly (A) addition and processing of hnRNA in Hela cells. *Cell*, **3**, 255–64.

Derman, E., Krauter, K., Walling, L., Weinberger, C. & Darnell, Jr, J. E. (1981). Transcriptional control in the production of liver-specific mRNAs. *Cell*, **23**, 731–9.

De Robertis, E. M. (1983). Nucleocytoplasmic segregation of proteins and RNAs. *Cell*, **32**, 1021–5.

De Robertis, E. M. & Olson, M. V. (1979). Transcription and processing cloned yeast tyrosine tRNA genes microinjected into frog oocytes. *Nature, Lond.* **278**, 137–41.

Diaz, M. O., Barsacchi-Pilone, G., Mahon, K. A. & Gall, J. G. (1981). Transcripts from both strands of a satellite DNA occur on lampbrush chromosome loops of the newt *Notophthalmus*. *Cell*, **24**, 649–59.

Dierks, P., van Ooyen, A., Mantei, N. & Wissman, C. (1981). DNA sequences preluding the rabbit β globin gene are required for formation in mouse L cells of β globin mRNA with the correct 5' terminus. *Proc. natn. Acad. Sci. USA*, **78**, 1411–15.

Dierks, P., van Ooyen, A., Cochran, M. D., Dobkin, C., Reiser, J. & Weissman, C. (1983). Three regions upstream from the cap site are required for efficient and accurate transcription of the rabbit β globin gene in mouse 3T6 cells. *Cell*, **32**, 695–706.

Diesseroth, A., Bode, U., Fontana, J. & Hendrick, D. (1980). Expression of human α globin genes in hybrid mouse erythroleukemia cells depends on differentiated state of human donor cell. *Nature, Lond.* **285**, 36–8.

Diesseroth, A., Nienhuis, A., Turner, P., Velez, R., Anderson, W. F., Ruddle, F., Lawrence, J., Creagan, R. & Kucherlapati, R. (1977). Localisation of the human α globin structural gene to chromosome 16 in somatic cell hybrids by molecular hybridisation assay. *Cell*, **12**, 205–18.

Diesseroth, A., Nienhuis, A., Lawrence, J., Giles, R., Turner, P. & Ruddle, F. H. (1978). Chromosomal localisation of human β globin gene on human chromosome 11 in somatic cell hybrids. *Proc. natn., Acad. Sci. USA*, **75**, 1456–60.

Dodemont, H. J., Soriano, P., Quax, W. J., Ramaekers, F., Lenstra, J. A., Groenen, M. A. M., Bernardi, G. & Bloemendahl, H. (1982). The genes coding for the cytoskeletal proteins actin and vimentin in warm-blooded vertebrates. *EMBO J.* **1**, 167–71.

Dolan, M., Sugarman, B. M., Dodgson, J. B. & Engel, J. D. (1981). Chromosomal arrangement of the chicken β type globin genes. *Cell*, **24**, 669–77.

Donohoo, P. & Kafatos, F. C. (1973). Differences in the proteins synthesised by the progeny of the first two blastomeres of *Ilyanassa*, a "mosaic" embryo. *Devl Biol.* **32**, 224–8.

Doolittle, W. F. (1982). Selfish DNA after fourteen months. In *Genome Evolution*, ed. G. A. Dover & R. B. Flavell, pp. 3–28. Academic Press, New York & London.

Driesch, H. (1898). Uber rein-mütterliche charaktere und bastardlaven von Echiniden. *Arch. EntwMech. Org.* **7**, 65–102.

Dugaiczyk, A., Woo, S. L. C., Colbert, D. A., Lai, E. C., Mace, Jr, M. L. & O'Malley, B. W. (1979). The ovalbumin gene: cloning and molecular organisation of the entire natural gene. *Proc. natn. Acad. Sci. USA*, **76**, 2253–7.

Dworniczak, B., Seidel, R. & Pongs, O. (1983). Puffing activities and binding of ecdysteroid to polytene chromosomes of *Drosophila melanogaster*. *EMBO J.* **2**, 1323–30.

Dynan, W. S. & Tijan, R. (1983). Isolation of transcription factors that discriminate between different promoters recognised by RNA polymerase 2. *Cell*, **32**, 669–80.

Early, P., Huang, H., Davis, M., Calame, K. & Hood, L. (1980a). An immunoglobulin heavy chain variable region is generated from three segments of DNA; V_H, D and J_H. *Cell*, **19**, 981–92.

Early, P., Rogers, J., Davis, M., Calame, K., Bond, M., Wall, R. & Hood, L. (1980b). Two mRNAs can be produced from a single immunoglobulin μ gene by alternative RNA processing pathways. *Cell*, **20**, 313–19.

Ede, D. A. (1978). *An Introduction to Developmental Biology*. Blackie, London & Glasgow.

Edwards, J. S., Milner, M. J. & Chen, S. W. (1978). Integument and sensory nerve differentiation of *Drosophila* leg and wing imaginal discs *in vitro*. *Roux' Arch.* **185**, 59–77.

Efstratiadis, A., Posakony, J. W., Maniatis, T., Lawn, R. M., O'Connell, C., Spritz, R. A., de Riel, J. K., Forget, B. G., Weissmann, S. M., Slightom, J. L., Blechl, A. E., Smithies, O., Baralle, F. E., Shoulders, C. C. & Proudfoot, N. J. (1980). The structure and evolution of the human β globin gene cluster. *Cell*, **21**, 653–68.

Elgin, S. C. R. (1982). Chromatin structure, DNA structure. *Nature, Lond.* **300**, 402–3.

Engel, J. D. & Dodgson, J. B. (1981). Histone genes are clustered but not tandemly repeated in the chicken genome. *Proc. natn. Acad. Sci. USA*, **78**, 2856–60.

Engelke, D. R., Ng, S.-Y., Shastry, B. S. & Roeder, R. G. (1980). Specific interaction of a purified transcription factor with an internal control region of 5S RNA genes. *Cell*, **19**, 717–28.

Ernst, S. G., Britten, R. J. & Davidson, E. H. (1979). Distinct single copy sequence sets in sea urchin nuclear RNAs. *Proc. natn. Acad. Sci. USA*, **76**, 2209–12.

Ernst, S. G., Hough-Evans, B. R., Britten, R. J. & Davidson, E. H. (1980). Limited complexity of the RNA in micromeres of sixteen-cell sea urchin embryos. *Devl Biol.* **79**, 119–27.

Fansler, B. & Loeb, L. A. (1969). Sea urchin DNA polymerase: II, changing localisation during early development. *Expl Cell Res.* **57**, 305–10.

Farrelly, F. & Butow, R. A. (1983). Rearranged mitochondrial genes in the yeast nuclear genome. *Nature, Lond.* **301**, 396–9.

Fedoroff, N. V. & Brown, D. D. (1978). The nucleotide sequence of oocyte 5S DNA in *X. laevis:* I, the AT-rich spacer. *Cell*, **13**, 701–16.

Felsenfeld, G. & McGhee, J. (1982). Methylation and gene control. *Nature, Lond.* **296**, 602–3.

Finch, J. T., Noll, M. & Kornberg, R. D. (1975). Electron microscopy of defined lengths of chromatin. *Proc. natn. Acad. Sci. USA*, **72**, 3320–2.

Flavell, A. (1983). *Drosophila* takes off. *Nature, Lond.* **305**, 96–7.

Flavell, R. A. (1982). The mystery of the mouse α globin pseudogene. *Nature, Lond.* **295**, 370.

Flavell, A. J. & Ish-Horowicz, D. (1981). Extrachromosomal circular copies of the eukaryotic transposable element *copia* in cultured *Drosophila* cells. *Nature, Lond.* **292**, 591–4.

Flavell, A. & Ish-Horowicz, D. (1983). The origin of extrachromosomal circular *copia* elements. *Cell,* **34**, 415–19.

Foe, V. E. (1977). Modulation of ribosomal RNA synthesis in *Onocopeltus fasciatus:* an electron microscopic study of the relationship between changes in chromatin structure and transcriptional activity. *Cold Spring Harb. Symp. quant. Biol.* **42**, 723–39.

Folk, W. B. & Hofstetter, H. (1983). A detailed mutational analysis of the eucaryotic tRNA$_1^{Met}$ promoter. *Cell,* **33**, 585–93.

Francke, C., Edstrom, J. E., McDowall, A. W. & Miller, O. L. Jr (1982). Electron microscopic visualisation of a discrete class of giant translation units in salivary gland cells of *Chironomus tentans. EMBO J.* **1**, 59–62.

Freeman, G. (1976). The role of cleavage in the localisation of developmental potential in the ctenophore *Mnemiopsis leidyi. Devl Biol.* **49**, 143–77.

Freeman, G. & Reynolds, G. T. (1973). The development of bioluminescence in the ctenophore *Mnemiopsis leidyi. Devl Biol.* **31**, 61–100.

Fritsch, E. F., Lawn, R. M. & Maniatis, T. (1980). Molecular cloning and characterisation of the human β-like globin gene cluster. *Cell,* **19**, 959–72.

Fyrberg, E. A., Kindle, K. L., Davidson, N. & Sidja, A. (1980). The actin genes of *Drosophila:* a dispersed multigene family. *Cell,* **19**, 365–78.

Fyrberg, E. A., Mahaffery, J. W., Bond, B. J. & Davidson, N. (1983). Transcripts of the six *Drosophila* actin genes accumulate in a stage and tissue specific manner. *Cell,* **33**, 115–23.

Galau, G. A., Klein, W. H., Davis, M. M., Wold, B. J., Britten, R. J. & Davidson, E. H. (1976). Structural gene sets active in embryos and adult tissues of the sea urchin. *Cell,* **7**, 487–505.

Galau, G. A., Klein, W. H., Britten, R. J. & Davidson, E. H. (1977). Significance of rare mRNA sequences in liver. *Archs Biochem. Biophys.* **179**, 584–99.

Galli, G., Hofstetter, H. & Birnstiel, M. L. (1981). Two conserved sequence blocks within eucaryotic tRNA genes are major promoter elements. *Nature, Lond.* **294**, 626–9.

Gallo, R. C. (1971). Reverse transcriptase: the DNA polymerase of oncogenic RNA viruses. *Nature, Lond.* **234**, 194–8.

Garber, R. L., Kuroiwa, A. & Gehring, W. J. (1983). Genomic and cDNA clones of the homoeotic locus *Antennapedia* in *Drosophila. EMBO J.* **2**, 2027–36.

Garcia-Bellido, A. (1975). Genetic control of wing disc development in *Drosophila.* In *Cell Patterning, Ciba Foundation Symposium,* pp. 161–78. Associated Scientific Publishers, Amsterdam.

Garcia-Bellido, A., Lawrence, P. A. & Morata, G. (1979). Compartments in animal development. *Scientific American,* **241**, 102–10.

Garcia-Bellido, A. & Merriam, J. R. (1969). Cell lineage of the imaginal discs in *Drosophila* gynandromorphs. *J. exp. Zool.* **170**, 61–75.

Garcia-Bellido, A., Ripoll, P. & Morata, G. (1973). Developmental compartments of the wing disk of *Drosophila. Nature New Biol.* **245**, 251–3.

Garel, A. & Axel, R. (1976). Selective digestion of transcriptionally-active ovalbumin genes from oviduct nuclei. *Proc. natn. Acad. Sci. USA*, **73**, 3966–70.

Garel, A., Zolan, M. & Axel, R. (1977). Genes transcribed at diverse rates have a similar conformation in chromatin. *Proc. natn. Acad. Sci. USA*, **74**, 4867–71.

Gazit, B., Panet, A. & Cedar, H. (1980). Reconstitution of a deoxyribonuclease I-sensitive structure on active genes. *Proc. natn. Acad. Sci. USA*, **77**, 1787–90.

Gehring, W. (1972). The stability of the determined state in cultures of imaginal disks in Drosophila. In *The Biology of Imaginal Disks*, ed. H. Ursprung & R. Nothiger, *Results and Problems in Cell Differentiation*, vol. 5, pp. 37–57. Springer-Verlag, New York, Heidelberg, Berlin.

Georgiev, G. P. (1969). On the structural organisation of operon and the regulation of RNA synthesis in animal cells. *J. theor. Biol.* **25**, 473–90.

Gerber-Huber, S., May, F. E. B., Westley, B. R., Felber, B. K., Hosbach, H. A., Andres, A. C. & Ryffel, G. U. (1983). In contrast to other *Xenopus* genes the estrogen-inducible vitellogenin genes are expressed when totally methylated. *Cell*, **33**, 43–51.

Geyer-Duszynska, I. (1966). Genetic factors in oogenesis and spermatogenesis in *Cecidomyiidae*. *Chromosomes Today*, **1**, 174–8.

Ghysen, A., Dambly-Chaudière, C., Jan, L. Y. & Jan, Y. N. (1982). Segmental differences in the protein content of imaginal discs. *EMBO J.* **1**, 1373–9.

Gianni, A. M., Bregni, M., Cappellini, M. D., Fiorelli, G., Taramelli, R., Giglioni, B., Comi, P. & Ottolenghi, S. (1983). A gene controlling fetal hemoglobin expression in adults is not linked to the non-α-globin cluster. *EMBO J.* **2**, 921–5.

Giglioni, G., Gianni, A. M., Comi, P., Ottolenghi, S. & Runnger, D. (1973). Translational control of globin synthesis by haemin in *Xenopus* oocytes. *Nature New Biol.* **246**, 99–102.

Gilbert, W. (1978). Why genes in pieces? *Nature, Lond.* **271**, 501.

Gillies, S. D., Morrison, S. L., Oi, V. T. & Tonegawa, S. (1983). A tissue-specific transcription enhancer element is located in the major intron of a rearranged immunoglobulin heavy-chain gene. *Cell*, **33**, 717–28.

Glover, D. M., Zaha, A., Stocker, A. J., Santelli, R. V., Pueyo, M. T., de Toledo, S. M. & Lara, F. J. S. (1982). Gene amplification in *Rhynchosciara* salivary gland chromosomes. *Proc. natn. Acad. Sci. USA*, **79**, 2947–51.

Golden, L., Schafer, U. & Rosbash, M. (1980). Accumulation of individual poly (A)$^+$ RNAs during oogenesis of *Xenopus laevis*. *Cell*, **22**, 835–44.

Goodbourn, S. E. Y., Higgs, D. R., Clegg, J. B. & Weatherall, D. J. (1983). Molecular basis of length polymorphism in the human ζ globin gene complex. *Proc. natn. Acad. Sci. USA*, **80**, 5022–6.

Gough, N. M., Kemp, D. J., Tyler, B. M., Adam, J. M. & Cory, S. (1980). Intervening sequences divide the gene for the constant region of mouse immunoglobulin μ chains into segments, each encoding a domain. *Proc. natn. Acad. Sci. USA*, **77**, 554–8.

Greenberg, R. M. & Adler, P. N. (1982). Protein synthesis and accumulation in *Drosophila melanogaster* imaginal discs: identification of a protein with a nonrandom spatial distribution. *Devl Biol.* **89**, 273–86.

Greenleaf, A. L., Borsett, L. M., Jiamachello, P. F. & Coulter, D. E. (1979). α-amanitin resistant *D. melanogaster* with an altered RNA polymerase II. *Cell*, **18**, 613–22.

Greenleaf, A. L., Weeks, J. R., Voelker, R. A., Ohnishi, S. & Dickson, B. (1980). Genetic and biochemical characterisation of mutants at an RNA polymerase II locus in *D. melanogaster*. *Cell*, **21**, 785–92.

Greenwald, I. S., Sternberg, P. W. & Horvitz, H. R. (1983). The *lin-12* locus specifies cell fates in *Caenorhabditis elegans*. *Cell*, **34**, 435–44.

Griffin-Shea, R., Thireos, G. & Kafatos, F. C. (1982). Organisation of a cluster of four chorion genes in *Drosophila* and its relationship to developmental expression and amplification. *Devl Biol.* **91**, 325–36.

Gronemeyer, H. & Pongs, O. (1980). Localisation of ecdysterone on polytene chromosomes of *Drosophila melanogaster*. *Proc. natn. Acad. Sci. USA*, **77**, 2108–12.

Grossbach, U. (1973). Chromosome puffs and gene expression in polytene cells. *Cold Spring Harb. Symp. quant. Biol.* **38**, 619–27.

Grosschedl, R. & Birnstiel, M. L. (1980a). Identification of regulatory sequences in the prelude sequences of an H2A histone gene by the study of specific deletion mutants *in vivo*. *Proc. natn. Acad. Sci. USA*, **77**, 1432–6.

Grosschedl, R. & Birnstiel, M. L. (1980b). Spacer DNA sequences upstream of the TATAAATA sequence are essential for promotion of H2A histone gene transcription *in vivo*. *Proc. natn. Acad. Sci. USA*, **77**, 7102–6.

Grosveld, G. C., Shewmaker, C. K., Jat, P. & Flavell, R. A. (1981a). Localisation of DNA sequences necessary for transcription of the rabbit β-globin gene *in vitro*. *Cell*, **25**, 215–26.

Grosveld, G. C., Koster, A. & Flavell, R. A. (1981b). A transcription map of the rabbit β-globin gene. *Cell*, **23**, 573–84.

Groudine, M., Peretz, M. & Weintraub, H. (1981). Transcriptional regulation of haemoglobin switching in chickens. *Mol. cell. Biol.* **1**, 281–8.

Groudine, M. & Weintraub, H. (1975). Rous sarcoma virus activates embryonic globin genes in chicken fibroblasts. *Proc. natn. Acad. Sci. USA*, **72**, 4464–8.

Groudine, M. & Weintraub, H. (1981). Activation of globin genes during chicken development. *Cell*, **24**, 393–401.

Groudine, M. & Weintraub, H. (1982). Propagation of globin DNase I-hypersensitive sites in the absence of factors required for induction: a possible mechanism for determination. *Cell*, **30**, 131–9.

Gruenbaum, Y., Cedar, H. & Razin, A. (1982). Substrate and sequence specificity of a eucaryotic DNA methylase. *Nature, Lond.* **295**, 620–2.

Grummt, I. (1981). Specific transcription of mouse ribosomal DNA in a cell-free system that mimics control *in vivo*. *Proc. natn. Acad. Sci. USA*, **78**, 727–31.

Grummt, I. & Lindigkeit, R. (1973). Preribosomal RNA synthesis in isolated rat-liver nucleoli. *Eur. J. Biochem.* **36**, 244–9.

Grummt, I., Roth, E. & Paule, M. R. (1982). Ribosomal RNA transcription *in vitro* is species-specific. *Nature, Lond.* **296**, 173–5.

Gurdon, J. B. (1962). Adult frogs derived from the nuclei of single somatic cells. *Devl Biol.* **4**, 256–73.

Gurdon, J. B., Lane, C. D., Woodland, H. R. & Marbaix, G. (1971). Use of frog eggs and oocytes for the study of messenger RNA and its translation in living cells. *Nature, Lond.* **233**, 177–81.

Gurdon, J. B. & Laskey, R. A. (1970). The transplantation of nuclei from single cultured cells into enucleated frogs' eggs. *J. Embryol. exp. Morph.* **24**, 227–48.

Gurdon, J. B. & Uehlinger, V. (1966). "Fertile" intestinal nuclei. *Nature, Lond.* **210**, 1240–1.

Hadorn, E. (1978). Transdetermination. In *The Genetics and Biology of Drosophila* ed. M. Ashburner & T. R. F. Wright, vol. 2c, pp. 555–617. Academic Press, New York & London.

Hamada, H., Muramatsu, M., Urano, Y., Onishi, T. & Kominami, R. (1979). *In vitro* synthesis of a 5S RNA precursor by isolated nuclei of rat liver and Hela cells. *Cell*, **17**, 163–73.

Hanas, J. S., Bogenhagen, D. F. & Wu, C. W. (1983). Cooperative model for the binding of *Xenopus* transcription factor A to the 5S gene. *Proc. natn. Acad. Sci. USA*, **80**, 2142–5.

Harkey, M. A. & Whiteley, A. H. (1983). The program of protein synthesis during the development of the micromere – primary mesenchyme cell line in the sea urchin embryo. *Devl Biol.* **100**, 12–28.

Harper, M. L. & Monk, M. (1983). Evidence for translation of HPRT enzyme on maternal mRNA in early mouse embryos. *J. Embryol. exp. Morph.* **74**, 15–28.

Harpold, M. M., Evans, R. M., Salditt-Georgieff, M. & Darnell Jr., J. E. (1979). Production of mRNA in Chinese hamster ovary cells: relationship of the rate of synthesis to the cytoplasmic concentration of nine specific mRNA sequences. *Cell*, **17**, 1025–35.

Harvey, E. B. (1936). Parthenogenetic merogony or cleavage without nuclei in *Arbacia punctulata. Biol. Bull.* **71**, 101–21.

Hayes, P. H., Sato, T. & Denell, R. E. (1984). Homoeosis in *Drosophila:* the *Ultrabithorax* larval syndrome. *Proc. natn. Acad. Sci. USA*, **81**, 545–9.

Heintz, N., Zernik, M. & Roeder, R. G. (1981). The structure of the human histone genes: clustered but not tandemly repeated. *Cell*, **24**, 661–8.

Hewish, D. R. & Burgoyne, L. A. (1973). Chromatin substructure: the digestion of chromatin DNA at regularly spaced sites by a nuclear DNase. *Biochem. biophys. Res. Commun.* **52**, 504–10.

Hill, R. J. & Stollar, D. B. (1983). Dependence of Z-DNA antibody binding to polytene chromosomes on acid fixation and DNA torsional strain. *Nature, Lond.* **305**, 338–40.

Hochmann, B. (1973). Analysis of a whole chromosome in *Drosophila. Cold Spring Harb. Symp. quant. Biol.* **38**, 581–9.

Hofstetter, H., Kressmann, A. & Birnstiel, M. L. (1981). A split promoter for a eucaryotic tRNA gene. *Cell*, **24**, 573–85.

Hollis, G. F., Hieter, P. A., McBride, O. W., Swan, D. & Leder, P. (1982). Processed genes: a dispersed human immunoglobulin gene bearing evidence of RNA-type processing. *Nature, Lond.* **296**, 321–6.

Honda, B. M. & Roeder, R. G. (1980). Association of a 5S gene transcription factor with 5S RNA and altered levels of the factor during cell differentiation. *Cell*, **22**, 119–26.

Hörstadius, S. (1928). Über die determination des keimes der Echinodermen. *Acta. zool. Stockh.* **9**, 1–192.

Hough-Evans, B. R., Ernst, S. G., Britten, R. J. & Davidson, E. H. (1979). RNA complexity in developing sea urchin oocytes. *Devl Biol.* **69**, 258–69.

Hough-Evans, B. R., Wold, B. J., Ernst, S. G., Britten, R. J. & Davidson, E. H. (1977). Appearance and persistence of maternal RNA sequences in sea urchin development. *Devl Biol.* **60**, 258–77.

Hourcade, D., Dressler, D. & Wolfson, J. (1973). The nucleolus and the rolling circle. *Cold Spring Harb. Symp. quant. Biol.* **38**, 537–50.

Hovemann, B., Sharp, S., Yamada, H. & Soll, D. (1980). Analysis of a *Drosophila* tRNA gene cluster. *Cell*, **19**, 889–95.

Huez, G., Marbaix, G., Hubert, E., Leclercq, M., Nudel, U., Soreq, H., Salomon, R., Lebleu, B., Revel, M. & Littauer, U. (1974). Role of the poly (A) segment in the translation of globin messenger RNA in *Xenopus* oocytes. *Proc. natn. Acad. Sci. USA*, **71**, 3143–6.

Humphries, S., Windass, J. & Williamson, R. (1976). Mouse globin gene expression in erythroid and non-erythroid tissues. *Cell*, **7**, 267–77.

Hyer, B. J. & Chan, L.-N. L. (1978). Initial synthesis of globin peptide chains in differentiating embryonic red blood cells. *Devl Biol.* **66**, 279–84.

Igo-Kemenes, T., Horz, W. & Zachau, H. G. (1982). Chromatin. *A. Rev. Biochem.* **51**, 89–121.

Illmensee, K. & Hoppe, P. C. (1981). Nuclear transplantation in *Mus musculus*: developmental potencies of nuclei from preimplantation embryos. *Cell*, **25**, 9–18.

Illmensee, K. & Mahowald, A. P. (1974). Transplantation of posterior pole plasm in *Drosophila:* induction of germ cells at the anterior pole of the egg. *Proc. natn. Acad. Sci. USA*, **71**, 1016–20.

Illmensee, K. (1972). Developmental potencies of nuclei from cleavage, preblastoderm and syncitial blastoderm transplanted into unfertilised eggs of *Drosophila melanogaster*. *Roux' Arch.* **170**, 267–98.

Ingham, P. W. (1983). Differential expression of bithorax complex genes in the absence of the *extra sex combs* and *trithorax* genes. *Nature, Lond.* **306**, 591–3.

Ingham, P. W. (1984). A gene that regulates the bithorax complex differentially in larval and adult cells of *Drosophila*. *Cell*, **37**, 815–23.

Ingles, C. J. (1978). Temperature-sensitive RNA polymerase II mutations in Chinese hamster ovary cells. *Proc. natn. Acad. Sci. USA*, **75**, 405–9.

Jäckle, H. & Kalthoff, K. (1980). Synthesis of posterior indicator protein in normal embryos and double abdomens of *Smittia* sp. (Chironomidae, Diptera). *Proc. natn. Acad. Sci. USA*, **77**, 1700–4.

Jäckle, H. & Kalthoff, K. (1981). Proteins foretelling head or abdomen development in the embryo of *Smittia* spec. (Chironomidae, Diptera). *Devl Biol.* **85**, 287–98.

Jacq, C., Miller, J. R. & Brownlee, G. G. (1977). A pseudogene structure in 5S DNA of *X. laevis. Cell*, **12**, 109–20.

Jaeger, A. W. & Kuenzle, C. C. (1982). The chromatin repeat length of brain cortex and cerebellar neurons changes concomitant with terminal differentiation. *EMBO J.* **1**, 811–16.

Jamrich, M., Greenleaf, A. L. & Bautz, E. K. F. (1977a). Localisation of RNA polymerase in polytene chromosomes of *Drosophila. Proc. natn. Acad. Sci. USA*, **74**, 2079–83.

Jamrich, M., Greenleaf, A. L. & Bautz, F. A. & Bautz, E. K. F. (1977b). Functional organisation of polytene chromosomes. *Cold Spring Harb. Symp. quant. Biol.* **42**, 389–96.

Jeffery, W. R., Tomlinson, C. R. & Brodeur, R. D. (1983). Localization of actin messenger RNA during early ascidean development. *Devl Biol.* **99**, 408–17.

Jeffreys, A. J. (1982). Evolution of globin genes. In *Genome Evolution*, ed. G. A. Dover & R. B. Flavell, pp. 157–76. Academic Press, New York & London.

John, B. & Miklos, G. L. G. (1979). Functional aspects of satellite DNA and heterochromatin. *Int. Rev. Cytol.* **58**, 1–114.

John, H. A., Patrinou-Georgopoulous, M. & Jones, K. W. (1977). Detection of myosin heavy chain mRNA during myogenesis in tissue culture by *in vitro* and *in situ* hybridisation. *Cell*, **12**, 501–8.

Johnston, R. N., Beverley, S. M. & Schimke, R. T. (1983). Rapid spontaneous dihydrofolate gene amplification shown by fluorescence-activated cell sorting. *Proc. natn. Acad. Sci. USA*, **80**, 3711–15.

Judd, B. H. & Young, M. W. (1973). An examination of the one cistron: one chromomere concept. *Cold Spring Harb. Symp. quant. Biol.*, **38**, 573–9.

Jung, A., Sippel, A. E., Grez, M. & Schutz, G. (1980). Exons encode functional and structural units of chicken lysozyme. *Proc. natn. Acad. Sci. USA*, **77**, 5759–63.

Kaine, B. P., Gupta, R. & Woese, C. R. (1983). Putative introns in tRNA genes of prokaryotes. *Proc. natn. Acad. Sci. USA*, **80**, 3309–12.

Karin, M., Haslinger, A., Holtgreve, H., Richards, R. I., Krauter, P., Westphal, H. M. & Beato, M. (1984). Characterisation of DNA sequences through which cadmium and glucocorticoid hormones induce human metallothionein-II$_A$ gene. *Nature, Lond.* **308**, 513–19.

Karn, J., Brenner, S. & Barnett, L. (1983). Protein structural domains in the *Caenorhabditis elegans unc*-54 myosin heavy chain gene are not separated by introns. *Proc. natn. Acad. Sci. USA*, **80**, 4253–7.

Kaufman, R. J., Brown, P. C. & Schimke, R. T. (1979). Amplified dihydrofolate reductase genes in unstably methotrexate resistant cells are associated with double minute chromosomes. *Proc. natn. Acad. Sci. USA*, **76**, 5669–73.

Kedes, L. & Maxson, R. (1981). Histone gene organisation: paradigm lost. *Nature, Lond.* **294**, 11–12.

Keene, M. A., Corces, V., Lowenhaupt, K. & Elgin, S. C. R. (1981). DNase I-hypersensitive sites in *Drosophila* chromatin occur at the 5' ends of regions of transcription. *Proc. natn. Acad. Sci. USA*, **78**, 143–6.

Kemp, D. J. (1975). Unique and repetitive sequences in multiple genes for feather keratin. *Nature, Lond.* **254**, 573–7.

King, C. R. & Piatigorsky, J. (1983). Alternative RNA splicing of the murine αA crystallin gene: protein coding information within an intron. *Cell*, **32**, 707–12.

King, R. C. & Aggarwal, S. K. (1965). Oogenesis in *Hyalophora cecropia*. *Growth*, **29**, 17–83.

King, W. J. & Greene, G. L. (1984). Monoclonal antibodies localize oestrogen receptor in the nuclei of target cells. *Nature, Lond.* **307**, 745–7.

Kleene, K. C. & Humphries, T. (1977). Similarity of hnRNA sequences in blastula and pluteus stage sea urchin embryos. *Cell*, **12**, 143–55.

Klein, W. H., Murphy, W., Attardi, G., Britten, R. J. & Davidson, E. H. (1974). Distribution of repetitive and nonrepetitive sequence transcripts in Hela mRNA. *Proc. natn. Acad. Sci. USA*, **71**, 1785–9.

Klug, W. S., King, R. C. & Wattiaux, J. M. (1970). Oogenesis in the *suppresor*[2] of *hairy-wing* mutant of *Drosophila melanogaster:* II nucleolar morphology and *in vitro* studies of RNA protein synthesis. *J. exp. Zool.* **174**, 125–40.

Kohorn, B. D. & Rae, P. M. M. (1983a). Localisation of DNA sequences promoting RNA polymerase I activity in *Drosophila*. *Proc. natn. Acad. Sci. USA*, **80**, 3265–8.

Kohorn, B. D. & Rae, P. M. M. (1983b). A component of *Drosophila* RNA polymerase I promoter lies within the rRNA transcription unit. *Nature, Lond.* **304**, 179–81.

Korn, L. J. & Brown, D. D. (1978). Nucleotide sequence of *Xenopus borealis* oocyte 5S DNA: comparison of sequences that flank several related eucaryotic genes. *Cell*, **15**, 1145–8.

Kornberg, R. D. (1974). Chromatin structure: a repeating unit of histones and DNA. *Science*, **184**, 868–71.

Kornberg, R. D. (1977). Structure of chromatin. *A. Rev. Biochem.* **46**, 931–54.

Kornberg, T. (1981). *engrailed*, a gene controlling compartment and segment formation in *Drosophila*. *Proc. natn. Acad. Sci. USA*, **78**, 1095–9.

Kornberg, R. D. & Thomas, J. O. (1974). Chromatin structure: oligomers of the histones. *Science*, **184**, 864–8.

Kraminsky, G. P., Clark, W. C., Estelle, M. A., Gietz, R. D., Sage, B. D., O'Connor, J. D. & Hodgetts, R. B. (1980). Induction of translatable mRNA for dopa decarboxylase in *Drosophila:* an early response to ecdysone. *Proc. natn. Acad. Sci. USA*, **77**, 4175–9.

Küntzel, H. & Kochel, H. G. (1981). Evolution of rRNA and origin of mitochondria. *Nature, Lond.* **293**, 751–3.

Labhart, P. & Reeder, R. H. (1984). Enhancer-like properties of the 60/81 bp elements in the ribosomal gene spacer of *Xenopus laevis*. *Cell*, **37**, 285–9.

Lacy, E. & Maniatis, T. (1980). The nucleotide sequence of a rabbit β globin pseudogene. *Cell*, **21**, 545–53.

Laemmli, U. K., Cheng, S. M., Adolph, K. W., Paulson, J. R., Brown, J. A. & Baumbach, W. R. (1977). Metaphase chromosome structure; the role of nonhistone proteins. *Cold Spring Harb. Symp. quant. Biol.* **42**, 351–60.

Lafer, E. M., Moller, A., Nordheim, A., Stollar, B. D. & Rich, A. (1981). Antibodies specific for left-handed Z-DNA. *Proc. natn. Acad. Sci. USA*, **78**, 3546–50.

Lai, E. C., Woo, S. L. C., Bordelon-Riser, M. E., Fraser, T. H. & O'Malley, B. W. (1980). Ovalbumin is synthesised in mouse cells transformed with the natural chicken ovalbumin gene. *Proc. natn. Acad. Sci. USA*, **77**, 244–8.

Lamb, M. M. & Daneholt, B. (1979). Characterisation of active transcription units in Balbiani rings of *Chironomus tentans*. *Cell*, **17**, 835–48.

Lane, C. D., Gurdon, J. B. & Woodland, H. R. (1974). Control of translation of globin mRNA in embryonic cells. *Nature, Lond.* **25**, 436–7.

Larsen, A. & Weintraub, H. (1982). An altered DNA conformation detected by S1 nuclease occurs at specific regions in active chick globin chromatin. *Cell*, **29**, 609–22.

Lauer, J., Shen, C. K. J. & Maniatis, T. (1980). The chromosomal arrangement of human α-like globin genes; sequence homology and α globin gene deletions. *Cell*, **20**, 119–30.

Laufer, J. S., Bazzicalupo, P. & Wood, W. B. (1980). Segregation of developmental potential in early embryos of *Caenorhabditis elegans*. *Cell*, **19**, 569–77.

Laughon, A. & Scott, M. P. (1984). Sequence of a *Drosophila* segmentation gene: protein structure homology with DNA-binding proteins. *Nature Lond.* **310**, 25–8.

Lawn, R. M., Efstratiadis, A., O'Connell, C. & Maniatis, T. (1980). The nucleotide sequence of the human β globin gene. *Cell*, **21**, 647–51.

Lawrence, P. A. (1981a). The cellular basis of segmentation in insects. *Cell*, **26**, 3–10.

Lawrence, P. A. (1981b). A general cell marker for clonal analysis of *Drosophila* development. *J. Embryol. exp. Morph.* **64**, 321–32.

Lawrence, P. A. & Brower, D. L. (1982). Myoblasts from *Drosophila* wing disks can contribute to developing muscles throughout the fly. *Nature, Lond.* **295**, 55–7.

Lawrence, P. A. & Morata, G. (1976). Compartments in the wing of *Drosophila*: a study of the *engrailed* gene. *Devl Biol.*, **50**, 321–37.

Lawrence, P. A. & Morata, G. (1983). The elements of the bithorax complex. *Cell*, **35**, 595–601.

Lawrence, P. A. & Struhl, G. (1982). Further studies of the *engrailed* phenotype in *Drosophila. EMBO J.* **1**, 827–33.

Lawson, G. M., Tsai, M. J. & O'Malley, B. W. (1980). Deoxyribonuclease I sensitivity of the nontranscribed sequences flanking the 5′ and 3′ of the ovomucoid gene and the ovalbumin and its related X and Y genes in hen oviduct nuclei. *Biochemistry*, **19**, 4403–11.

Leder, A., Swan, D., Ruddle, F., D'Eustachio, P. & Leder, P. (1981). Dispersion of α-like globin genes of the mouse to three different chromosomes. *Nature, Lond.* **293**, 196–200.

LeMeur, M., Glanville, N., Mandel, J. L., Gerlinger, P., Palmiter, R. & Chambon, P. (1981). The ovalbumin gene family: hormonal control of X and Y gene transcription and mRNA accumulation. *Cell*, **23**, 561–7.

Lerner, M. R., Boyle, J. A., Mount, S. M., Wolin, S. L. & Steitz, J. A. (1980). Are snRNPs involved in splicing? *Nature, Lond.* **283**, 220–4.

Levine, M., Garen, A., Lepesant, J. A. & Lepesant-Kejzlarova, J. (1981). Constancy of somatic DNA organisation in developmentally-regulated regions of the *Drosophila* genome. *Proc. natn. Acad. Sci. USA*, **78**, 2417–21.

Levine, M., Hafen, E., Garber, R. L. & Gehring, W. J. (1983). Spatial distribution of *Antennapedia* transcripts during *Drosophila* development. *EMBO J.* **2**, 2037–46.

Levitt, A., Axel, R. & Cedar, H. (1979). Nick translation of active genes in intact nuclei. *Devl Biol.* **69**, 496–505.

Lewin, B. (1974). *Gene Expression*, vol. I, *Procaryotic Genomes*. Wiley-Interscience, New York.

Lewin, B. (1977). *Gene Expression*, vol. III, *Plasmids and Phages*. Wiley-Interscience, New York.

Lewin, B. (1980). *Gene Expression*, vol. II, *Eucaryotic Chromosomes*, 2nd ed. Wiley-Interscience, New York.

Lewis, E. B. (1978). A gene complex controlling segmentation in *Drosophila. Nature, Lond.* **276**, 565–70.

Lifton, R. P., Goldberg, M. L., Karp, R. W. & Hogness, D. S. (1977). The organisation of the histone genes in *D. melanogaster*: functional and evolutionary implications. *Cold Spring Harb. Symp. quant. Biol.* **42**, 1047–51.

Lifton, R. P. & Kedes, L. H. (1976). Size and sequence homology of masked maternal and embryonic histone messenger RNAs. *Devl Biol.* **48**, 47–55.

Lindenmaier, W., Nguyen-Huu, M. C., Lurz, R., Stratmann, M., Blin, N., Wurtz, T., Hauser, H. J., Sippel, A. E. & Schutz, G. (1979). Arrangement of coding and intervening sequences in the chicken lysozyme gene. *Proc. natn. Acad. Sci. USA*, **76**, 6196–200.

Little, P. F. R. (1982). Globin pseudogenes. *Cell*, **28**, 683–4.

Loidl, P., Loidl, A., Puschendorf, B. & Grobner, P. (1983). Lack of correlation between histone H4 acetylation and transcription during the *Physarum* cell cycle. *Nature, Lond.* **305**, 446–8.

Luse, D. S. & Roeder, R. G. (1980). Accurate transcription initiation on a purified mouse globin gene fragment in a cell-free system. *Cell*, **20**, 691–9.

McCarthy, B. J. & Church, R. B. (1970). The specificity of molecular hybridisation reactions. *A. Rev. Biochem.* **39**, 131–50.

McGinnis, W., Garber, R. L., Wirz, J., Kuroiwa, A. & Gehring, W. J. (1984b). A homologous protein-coding sequence in *Drosophila* homoeotic genes and its conservation in other metazoans. *Cell*, **37**, 403–8.

McGinnis, W., Levine, M. S., Hafen, E., Kuroiwa, A. & Gehring, W. J. (1984a). A conserved DNA sequence in homoeotic genes of the *Drosophila* Antennapedia and bithorax complexes. *Nature, Lond.* **308**, 428–33.

McKnight, G. S. (1978). The induction of ovalbumin and conalbumin mRNA by oestrogen and progesterone in chick oviduct explant cultures. *Cell*, **14**, 403–13.

McKnight, S. L., Bustin, M. & Miller, Jr, O. L. (1977). Electron microscopic analysis of chromosome metabolism in the *Drosophila melanogaster* embryo: anti-H3 and H2B antibodies recognise "active" chromatin. *Cold Spring Harb. Symp. quant. Biol.* **42**, 741–54.

Macleod, D. & Bird, A. (1983). Transcription in oocytes of highly methylated rDNA from *Xenopus laevis* sperm. *Nature, Lond.* **306**, 200–3.

McNaughton, M., Freeman, K. B. & Bishop, J. O. (1974). A precursor to haemoglobin mRNA in nuclei of immature red blood cells. *Cell*, **1**, 117–25.

McReynolds, L., O'Malley, B. W., Nisbet, A. D., Fothergill, J. E., Givol, D., Fields, S., Robertson, M. & Brownlee, G. G. (1978). Sequence of chicken ovalbumin mRNA. *Nature, Lond.* **273**, 723–8.

Manning, J. E., Schmid, C. W. & Davidson, N. (1975). Interspersion of repetitive and non-repetitive DNA sequences in the *Drosophila* genome. *Cell*, **4**, 41–56.

Marcu, K. B. & Cooper, M. D. (1982). New views of the immunoglobulin heavy chain switch. *Nature, Lond.* **298**, 327–8.

Marie, J., Simon, M. P., Dreyfus, J. C. & Kahn, A. (1981). One gene but two messenger RNAs encode liver L and red cell L' pyruvate kinase subunits. *Nature, Lond.* **292**, 70–2.

Marzluff Jr, W. F., Murphy, E. C. & Huang, R. C. C. (1974). Transcription of the genes for 5S ribosomal RNA and transfer RNA in isolated mouse myeloma cell nuclei. *Biochemistry*, **13**, 3689–96.

Marzluff Jr, W. F. & Huang, R. C. C. (1975). Chromatin-directed transcription of 5S and tRNA genes. *Proc. natn. Acad. Sci. USA*, **72**, 1082–6.

Mavilio, F., Giampaolo, A., Carè, A., Migliaccio, G., Calandrini, M., Russo, G., Pagliardi, G. L., Mastroberardino, G., Marinucci, M. & Peschle, C.

(1983). Molecular mechanisms of human hemoglobin switching: selective undermethylation and expression of globin genes in embryonic, fetal, and adult erythroblasts. *Proc. natn. Acad., Sci. USA*, **80**, 6907–11.

Maxam, A. M. & Gilbert, W. (1977). A new method for sequencing DNA. *Proc. natn. Acad. Sci. USA*, **74**, 560–4.

Melli, M., Whitfield, C., Rao, K. V., Richardson, M. & Bishop, J. O. (1971). DNA-RNA hybridisation in vast DNA excess. *Nature New Biol.* **231**, 8–12.

Melton, D. A., de Robertis, E. M. & Cortese, R. (1980). Order and intracellular location of the events involved in the maturation of a spliced tRNA. *Nature, Lond.* **284**, 143–8.

Metafora, S., Felicetti, L. & Gambino, R. (1971). The mechanism of protein synthesis activation after fertilization of sea urchin eggs. *Proc. natn. Acad. Sci. USA*, **68**, 600.

Meyerowitz, E. M. & Hogness, D. S. (1982). Molecular organisation of a *Drosophila* puff site that responds to ecdysone. *Cell*, **28**, 165–76.

Milbrandt, J. D., Heintz, N. H., White, W. C., Rothman, S. M. & Hamlin, J. I. (1981). Methotrexate-resistant Chinese hamster ovary cells have amplified a 135 kilobase-pair region that includes the dihydrofolate reductase gene. *Proc. natn. Acad. Sci. USA*, **78**, 6043–7.

Miller, O. L. Jr. & Beatty, B. R. (1969). Portrait of a gene. *J. Cell Physiol.* **74** (suppl. 1), 225–32.

Milner, R. J., Bloom, F. E., Lai, C., Lerner, R. A. & Sutcliffe, J. G. (1984). Brain-specific genes have identifier sequences in their introns. *Proc. natn. Acad. Sci. USA*, **81**, 713–17.

Minty, A. J., Alonso, S., Caravatti, M. & Buckingham, M. E. (1982). A fetal skeletal muscle actin mRNA in the mouse and its identity with cardiac actin mRNA. *Cell*, **30**, 185–92.

Molgaard, H. V. (1980). Assembly of immunoglobulin heavy chain genes. *Nature, Lond.* **286**, 657–9.

Molgaard, H. V., Perucho, M. & Ruiz-Carillo, A. (1980). Histone H5 messenger RNA is polyadenylated. *Nature, Lond.* **283**, 502–4.

Molloy, G. W., Jelinek, W., Salditt, M. & Darnell, Jr, J. E. (1974). Arrangement of specific oligonucleotides within poly (A) terminated hnRNA molecules. *Cell*, **1**, 43–52.

Monk, M. & Harper, M. (1978). X chromosome activity in preimplantation embryos from XX and XO mothers. *J. Embryol. exp. Morph.* **46**, 53–64.

Montoya, J., Gaines, G. L. & Attardi, G. (1983). The pattern of transcription of the human mitochondrial rRNA genes reveals two overlapping transcription units. *Cell*, **34**, 151–9.

Moon, R. T., Danilchik, M. V. & Hille, M. B. (1982). An assessment of the masked messenger hypothesis: sea urchin egg messenger ribonucleoprotein complexes are efficient templates for *in vitro* protein synthesis. *Devl Biol.* **93**, 389–402.

Morata, G. & Kerridge, S. (1981). Sequential functions of the bithorax complex of *Drosophila*. *Nature, Lond.* **290**, 778–81.

Morata, G., Kornberg, T. & Lawrence, P. A. (1983). The phenotype of *engrailed* mutations in the antenna of *Drosophila*. *Devl Biol.* **99**, 27–33.

Morata, G. & Lawrence, P. A. (1977). Homoeotic genes, compartments and cell determination in *Drosophila*. *Nature, Lond.* **265**, 211–16.

Morata, G. & Ripoll, P. (1975). *Minutes*: mutants of *Drosophila* autonomously affecting cell division rate. *Devl Biol.* **42**, 211–21.

Moreau, J., Marcaud, L., Maschat, F., Kejzlarova-Lepesant, J., Lepesant, J. A. & Scherrer, K. (1982). A + T rich linkers define functional domains in eucaryotic DNA. *Nature, Lond.* **295**, 260–3.

Moritz, K. B. & Roth, G. E. (1976). Complexity of germline and somatic DNA in *Ascaris*. *Nature, Lond.* **259**, 55–7.

Moss, T. (1982). Transcription of cloned *Xenopus laevis* ribosomal DNA microinjected into *Xenopus* oocytes, and the identification of an RNA polymerase I promoter. *Cell*, **30**, 835–42.

Mount, S. & Steitz, J. (1983). Lessons from mutant globins. *Nature, Lond.* **303**, 380–1.

Moyzis, R. K., Bonnet, J., Li, D. W. & Ts'o, P. O. P. (1981a). An alternative view of mammalian DNA sequence organisation: I, repetitive sequence interspersion in Syrian hamster DNA: a model system. *J. molec. Biol.* **153**, 841–70.

Moyzis, R. K., Bonnet, J., Li, D. W. & Ts'o, P. O. P. (1981b). An alternative view of mammalian DNA sequence organisation: II, short repetitive sequences are organised into scrambled tandem clusters in Syrian hamster DNA. *J. molec. Biol.* **153**, 871–96.

Mulvihill, E. R., LePennec, J. P. & Chambon, P. (1982). Chicken oviduct progesterone receptor: location of specific regions of high-affinity binding in cloned DNA fragments of hormone-responsive genes. *Cell*, **28**, 621–32.

Murphy, D., Brickell, P. M., Latchman, D. S., Willison, K. & Rigby, P. W. J. (1983). Transcripts regulated during normal embryonic development and oncogenic transformation share a repetitive element. *Cell*, **35**, 865–71.

Naveh-Many, T. & Cedar, H. (1981). Active gene sequences are undermethylated. *Proc. natn. Acad. Sci. USA*, **78**, 4246–50.

Nevins, J. R. (1983). The pathway of eucaryotic mRNA formation. *A. Rev. Biochem.* **52**, 441–66.

Ng, S. Y., Parker, C. S. & Roeder, R. G. (1979). Transcription of cloned *Xenopus* 5S RNA genes by *X. laevis* RNA polymerase III in reconstituted systems. *Proc. natn. Acad. Sci. USA*, **76**, 136–40.

Nguyen-Huu, M., Sippel, A. E., Hynes, N. E., Groner, B. & Schutz, G. (1978). Preferential transcription of the ovalbumin gene in isolated hen oviduct nuclei by RNA polymerase B. *Proc. natn. Acad. Sci. USA*, **75**, 686–90.

Nguyen-Huu, M. C., Stratmann, M., Groner, B., Wurtz, T., Land, H., Giesecke, K., Sippel, A. E. & Schutz, G. (1979). Chicken lysozyme gene contains several intervening sequences. *Proc. natn. Acad. Sci. USA*, **76**, 76–8.

Nienhuis, A. W. & Stamatoyannopoulos, G. (1978). Haemoglobin switching. *Cell*, **15**, 307–15.

Nishioka, Y., Leder, A. & Leder, P. (1980). Unusual α globin-like gene that has cleanly lost both globin intervening sequences. *Proc. natn. Acad. Sci. USA*, **77**, 2806–9.

Noll, M. (1974a). Subunit structure of chromatin. *Nature, Lond.* **251**, 249–51.

Noll, M. (1974b). Internal structure of the chromatin subunit. *Nucl. Acids Res.* **1**, 1573–8.

Noll, M. & Kornberg, R. D. (1977). Action of micrococcal nuclease on chromatin and the location of H1. *J. molec. Biol.* **109**, 393–404.

Nordheim, A., Pardue, M.-L., Lafer, E. M., Möller, A., Stollar, B. D. & Rich, A. (1981). Antibodies to left-handed Z-DNA bind to interband regions of *Drosophila* polytene chromosomes. *Nature, Lond.* **294**, 417–20.

Nordheim, A., Tesser, P., Azorin, F., Kwon, Y. H., Möller, A. & Rich, A. (1982). Isolation of *Drosophila* proteins that bind selectively to left-handed Z-DNA. *Proc. natn. Acad. Sci. USA*, **79**, 7729–33.

Nordstrom, J. L., Roop, D. R., Tsai, M. J. & O'Malley, B. W. (1979). Identification of potential ovomucoid mRNA precursors in chick oviduct nuclei. *Nature, Lond.* **278**, 328–33.

North, G. (1983). Cloning the genes that specify fruit flies. *Nature, Lond.* **303**, 134–6.

Nothiger, R. (1972). The larval development of imaginal disks. In *The Biology of Imaginal Disks*, ed. H. Ursprung & R. Nothiger, *Results and Problems in Cell Differentiation*, vol. 5, pp. 1–33. Springer-Verlag, New York, Heidelberg, Berlin.

Nunberg, J. H., Kaufman, R. J., Schimke, R. T., Urlaub, G. & Chasin, L. A. (1978). Amplified dihydrofolate reductase genes are localised to a homogeneously-staining region of a single chromosome in a methotrexate-resistant Chinese hamster ovary cell line. *Proc. natn. Acad. Sci. USA*, **75**, 5553–6.

Nüsslein-Volhard, C. & Wieschaus, E. (1980). Mutations affecting segment number and polarity in *Drosophila*. *Nature, Lond.* **287**, 795–9.

O'Hare, K. & Rubin, G. M. (1983). Structures of P transposable elements and their sites of insertion and excision in the *Drosophila melanogaster* genome. *Cell*, **34**, 25–35.

Okazaki, K. (1975). Spicule formation by isolated micromeres of the sea urchin embryo. *Am. Zool.* **15**, 567–81.

Olins, A. L. & Olins, D. E. (1974). Spheroid chromatin units (ν bodies). *Science*, **183**, 330–2.

Orkin, S. H., Kazazian, Jr, H. H., Antonarakis, S. E., Ostrer, H., Goff, S. C. & Sexton, J. P. (1982). Abnormal RNA processing due to the exon mutation of β^E-globin gene. *Nature, Lond.* **300**, 768–70.

Osheim, Y. N. & Miller, Jr, O. L. (1983). Novel amplification and transcriptional activity of chorion genes in *Drosophila melanogaster* follicle cells. *Cell*, **33**, 543–53.

Ott, M. C., Sperling, L., Cassio, D., Levilliers, J., Sala-Trepat, J. & Weiss, M. C. (1982). Undermethylation at the 5' end of the albumin gene is necessary but not sufficient for albumin production by rat hepatoma cells in culture. *Cell*, **30**, 825–33.

Oudet, P., Germond, J. E., Bellard, M., Spadafora, C. & Chambon, P. (1978). Nucleosome structure. *Phil. Trans. R. Soc. Ser. B*, **283**, 241–58.

Oudet, P., Gros-Bellard, M. & Chambon, P. (1975). Electron microscopic and biochemical evidence that chromatin structure is a repeating unit. *Cell*, **4**, 281–300.

Padgett, R. A., Mount, S. M., Steitz, J. A. & Sharp, P. A. (1983). Splicing of messenger RNA precursors is inhibited by antisera to small nuclear ribonucleoprotein. *Cell*, **35**, 101–7.

Palmiter, R. D. (1975). Quantitation of the parameters that determine the rate of ovalbumin synthesis. *Cell*, **4**, 189–97.

Palmiter, R. D. & Carey, N. H. (1974). Rapid inactivation of ovalbumin messenger ribonucleic acid after acute withdrawal of estrogen. *Proc. natn. Acad. Sci. USA*, **71**, 2357–61.

Palmiter, R. D., Moore, P. B., Mulvihill, E. R. & Emtage, S. (1976). A significant lag in the induction of ovalbumin messenger RNA by steroid hormones: a receptor translocation hypothesis. *Cell*, **8**, 557–72.

Palmiter, R. D., Mulvihill, E. R., McKnight, G. S. & Senear, A. W. (1977). Regulation of gene expression in the chick oviduct by steroid hormones. *Cold Spring Harb. Symp. quant. Biol.* **42**, 639–47.

Pankow, W., Lezzi, M. & Holdregger-Mahling, I. (1976). Correlated changes of Balbiani ring expression and secretory protein synthesis in larval salivary glands of *Chironomus tentans. Chromosoma*, **58**, 137–53.

Parker, P. (1983). Enhancer elements activated by steroid hormones. *Nature, Lond.* **304**, 687–8.

Parker, C. S., Jaehning, J. A. & Roeder, R. G. (1977). Faithful gene transcription by eukaryotic RNA polymerases in reconstructed systems. *Cold Spring Harb. Symp. quant. Biol.* **42**, 577–87.

Parker, C. S. & Roeder, R. G. (1977). Selective and accurate transcription of the *Xenopus laevis* 5S RNA genes in isolated chromatin by purified RNA polymerase III. *Proc. natn. Acad. Sci. USA*, **74**, 44–8.

Paul, J. (1982). Gene switching and cellular differentiation. In *Stability and Switching in Cellular Differentiation*, ed. R. M. Clayton & D. E. S. Truman, *Adv. exp. Med. Biol.* vol. 158, pp. 13–21. Plenum Press, New York & London.

Pelham, H. R. B. & Brown, D. D. (1980). A specific transcription factor that can bind either the 5S RNA gene or 5S RNA. *Proc. natn. Acad. Sci. USA*, **77**, 4170–4.

Pelham, H. R. B., Wormington, W. M. & Brown, D. D. (1981) Related 5S transcription factors in *Xenopus* oocytes and somatic cells. *Proc. natn. Acad. Sci. USA*, **78**, 1760–4.

Perkowska, E., MacGregor, H. C. & Birnstiel, M. L. (1968). Gene amplification in the oocyte nucleus of mutant and wild-type *Xenopus laevis. Nature, Lond.* **217**, 649–50.

Pestell, R. Q. W. (1975). Microtubule protein synthesis during oogenesis and early embryogenesis in *Xenopus laevis. Biochem. J.* **145**, 527–34.

Picard, D. & Schaffner, W. (1984). A lymphocyte-specific enhancer in the mouse immunoglobulin κ gene. *Nature, Lond.* **307**, 80–3.

Posakony, J. W., Flytzanis, C. N., Britten, R. J. & Davidson, E. H. (1983). Interspersed sequence organisation and developmental representation of cloned poly (A)$^+$ RNAs from sea urchin eggs. *J. molec. Biol.* **167**, 361–89.

Potter, S. S., Brorein, Jr, W. J., Dunsmuir, P. & Rubin, G. M. (1979). Transposition of elements of the 412, *copia* and 297 dispersed repeated gene families in *Drosophila. Cell*, **17**, 415–27.

Prior, C. P., Cantor, C. R., Johnson, E. M., Littau, V. C. & Allfrey, V. G. (1983). Reversible changes in nucleosome structure and histone H3 accessibility in transcriptionally active and inactive states of rDNA chromatin. *Cell*, **34**, 1033–42.

Proudfoot, N. J., Gil, A. & Maniatis, T. (1982). The structure of the human zeta globin gene and a closely-linked, nearly identical pseudogene. *Cell*, **31**, 553–63.

Proudfoot, N. J. & Maniatis, T. (1980). The structure of a human α globin pseudogene and its relationship to α globin gene duplication. *Cell*, **21**, 537–44.

Quax, W., Egberts, W. V., Hendriks, W., Quax-Jeuken, Y. & Bloemendal, H. (1983). The structure of the vimentin gene. *Cell*, **35**, 215–23.

Queen, C. & Baltimore, D. (1983). Immunoglobulin gene transcription is activated by downstream sequence elements. *Cell*, **33**, 741–8.

Raff, R. A., Greenhouse, G., Gross, K. W. & Gross, P. R. (1971). Synthesis and storage of microtubule proteins by sea urchin embryos. *J. Cell Biol.* **50**, 516–27.

Raven, C. P. (1961). *Oogenesis: The Storage of Developmental Information.* Pergamon Press, Oxford.

Raven, C. P. (1963). Mechanisms of determination in the development of gastropods. *Adv. Morphogen.* **3**, 1–32.

Reeder, R. H. & Roeder, R. G. (1972). Ribosomal RNA synthesis in isolated nuclei. *J. molec. Biol.* **67**, 433–41.

Renkawitz, R., Beug, H., Graf, T., Matthias, P., Grez, M. & Schutz, G. (1982). Expression of a chicken lysozyme recombinant gene is regulated by progesterone and dexamethasone after microinjection into oviduct cells. *Cell*, **31**, 167–76.

Roberts, J. M., Buck, L. B. & Axel, R. (1983). A structure for amplified DNA. *Cell*, **33**, 53–63.

Rodgers, W. H. & Gross, P. R. (1978). Inhomogeneous distribution of egg RNA sequences in the early embryo. *Cell*, **14**, 279–88.

Roeder, R. G. (1974). Multiple forms of deoxyribonucleic acid-dependent ribonucleic acid polymerase in *Xenopus laevis:* levels of activity during oocyte and embryonic development. *J. biol. Chem.* **249**, 249–56.

Roeder, R. G. & Rutter, W. J. (1969). Multiple forms of DNA-dependent RNA polymerase in eukaryotic organisms. *Nature, Lond.* **224**, 234–7.

Roeder, R. G. & Rutter, W. J. (1970). Multiple RNA polymerases and RNA synthesis during sea urchin development. *Biochemistry*, **9**, 2543–53.

Rogers, J., Early, P., Carter, C., Calame, K., Hood, L. & Wall, R. (1980). Two mRNAs with different 3′ ends encode membrane-bound and secreted forms of immunoglobulin μ chain. *Cell*, **20**, 303–12.

Rogers, J. & Wall, R. (1980). A mechanism for RNA splicing. *Proc. natn. Acad. Sci. USA*, **77**, 1877–9.

Roop, D. R., Kristo, P., Stumph, W. E., Tsai, M. J. & O'Malley, B. W. (1981). Structure and expression of a chicken gene coding for UI RNA. *Cell*, **23**, 671–80.

Rosbash, M. & Ford, P. J. (1974). Polyadenylic acid-containing RNA in *Xenopus laevis* oocytes. *J. molec. Biol.* **85**, 87–101.

Ross, J. (1976). A precursor of globin messenger RNA. *J. molec. Biol.* **106**, 403–20.

Royal, A., Garapin, A., Cami, B., Perrin, F., Mandel, J. L., LeMeur, M., Brégegère, F., Gannon, F., LePennec, J. P., Chambon, P. & Kourilsky, P. (1979). The ovalbumin gene region: common features in the organisation of three genes expressed in chick oviduct under hormonal control. *Nature, Lond.* **279**, 125–32.

Rubin, G. M. & Hogness, D. S. (1975). Effect of heat shock on the synthesis of low molecular weight RNAs in *Drosophila:* accumulation of a novel form of 5S RNA. *Cell*, **6**, 207–16.

Runnström, J. (1928). Plasmabau und determination bei dem ei von *Paracentrotus lividus*, LK. *Roux'Arch.* **113**, 556–81.

Ryffel, G. U., Muellener, D. B., Wyler, T., Wahli, W. & Weber, R. (1981). Transcription of single copy vitellogenin gene of *Xenopus* involves expression of middle repetitive DNA. *Nature, Lond.* **291**, 429–30.

Ryffel, G. U., Muellener, D. B., Gerber-Huber, S., Wyler, T. & Wahli, W. (1983). Scattering of repetitive DNA sequences in the albumin and vitellogenin gene loci of *Xenopus laevis*. *Nucl. Acids Res.* **11**, 7701–16.

Sakano, H., Rogers, J. H., Huppi, K., Brack, G., Traunecker, A., Maki, R., Wall, R. & Tonegawa, S. (1979). Domains and the hinge region of an immunoglobulin heavy chain are encoded in separate DNA segments. *Nature, Lond.* **277**, 627–31.

Sakonju, S., Bogenhagen, D. F. & Brown, D. D. (1980). A control region in the centre of the 5S gene directs specific initiation of transcription: I, the 5' border of the region. *Cell*, **19**, 13–25.

Salditt-Georgieff, M. & Darnell Jr, J. E. (1982). Further evidence that the majority of primary nuclear RNA transcripts in mammalian cells do not contribute to mRNA. *Mol. cell. Biol.* **2**, 701–7.

Salditt-Georgieff, M., Harpold, M., Sawicki, S., Nevins, J. & Darnell Jr, J. E. (1980). Addition of poly (A) to nuclear RNA occurs soon after RNA synthesis. *J. Cell Biol.* **86**, 844–8.

Samal, B., Worcel, A., Louis, C. & Schedl, A. (1981). Chromatin structure of the histone genes of *D. melanogaster*. *Cell*, **23**, 401–9.

Sanger, F., Nicklen, S. & Coulson, A. R. (1977). DNA sequencing with chain-terminating inhibitors. *Proc. natn. Acad. Sci. USA*, **74**, 5463–7.

Sawicki, J. A., Magnuson, T. & Epstein, C. J. (1981). Evidence for expression of the paternal genome in the two-cell mouse embryo. *Nature, Lond.* **294**, 450–2.

Schaltmann, K. & Pongs, O. (1982). Identification and characterisation of the ecdysterone receptor in *Drosophila melanogaster* by photoaffinity labelling. *Proc. natn. Acad. Sci. USA*, **79**, 6–10.

Schechter, I. (1974). Use of antibodies for the isolation of biologically pure messenger ribonucleic acid from fully functional eukaryotic cells. *Biochemistry*, **13**, 1875–85.

Scheidereit, C., Geisse, S., Westphal, H. M. & Beato, M. (1983). The glucocorticoid receptor binds to defined nucleotide sequences near the promoter in mouse mammary tumor virus. *Nature, Lond.* **304**. 749–51.

Scheller, R. H., Costantini, F. D., Kozlowski, M. R., Britten, R. J. & Davidson, E. H. (1978). Specific representation of cloned repetitive DNA sequences in sea urchin RNAs. *Cell*, **15**, 189–203.

Schibler, U., Hagenbüchle, O., Young, R. A., Tomi, M. & Wellauer, P. K. (1982). Tissue-specific expression of mouse α-amylase genes. In *Stability and Switching in Cellular Differentiation*, ed. R. M. Clayton & D. E. S. Truman, *Adv. exp. Med. Biol.*, vol. 158, pp. 381–5. Plenum Press, New York & London.

Schibler, U., Hagenbüchle, O., Wellauer, P. K. & Pittet, A. C. (1983). Two promoters of different strengths control the transcription of the mouse alpha-amylase gene *Amy* 1^a in the parotid gland and the liver. *Cell*, **33**, 501–8.

Schmid, C. W. & Deininger, P. L. (1975). Sequence organisation of the human genome. *Cell*, **6**, 345–58.

Scholnick, S. B., Morgan, B. A. & Hirsch, J. (1983). The cloned dopa decarboxylase gene is developmentally regulated when reintegrated into the *Drosophila* genome. *Cell*, **34**, 37–45.

Schubiger, G. (1971). Regeneration, pattern duplication and transdetermination in fragments of the leg disc of *Drosophila melanogaster*. *Devl Biol.* **26**, 277–95.

Schutz, G., Nguyen-Huu, M. C., Giesecke, K., Hynes, N. E., Groner, B., Wurtz, T. & Sippel, A. E. (1977). Hormonal control of egg white protein messenger RNA synthesis in the chicken oviduct. *Cold Spring Harb. Symp. quant. Biol.* **42**, 617–24.

Scott, S. E. M. & Sommerville, J. (1974). Location of nuclear proteins on the chromosomes of newt oocytes. *Nature, Lond.* **250**, 680–2.

Scott, M. P., Weiner, A. J., Hazelrigg, T. I., Polisky, B. A., Pirotta, V., Scalenghe, F. & Kaufman, T. C. (1983). The molecular organisation of the *Antennapedia* locus of *Drosophila*. *Cell*, **35**, 763–76.

Searle, P. F. & Tata, J. R. (1981). Vitellogenin gene expression in male *Xenopus* hepatocytes during primary and secondary stimulation with estrogen in cell culture. *Cell*, **23**, 741–6.

Senger, D. R., Arceci, R. J. & Gross, P. R. (1978). Histones of sea urchin embryos: transients in transcription, translation and composition of chromatin. *Devl Biol.* **65**, 416–25.

Senger, D. R. & Gross, P. R. (1978). Macromolecule synthesis and determination in sea urchin blastomeres at the sixteen-cell stage. *Devl Biol.* **65**, 404–15.

Sharp, S., de Franco, D., Dingermann, T., Farrell, P. & Soll, D. (1981). Internal control regions for transcription of eucaryotic tRNA genes. *Proc. natn. Acad. Sci. USA*, **78**, 6657–61.

Shen, D. W., Real, F. X., DeLeo, A. B., Old, L. J., Marks, P. A. & Rifkind, R. A. (1983). Protein p53 and inducer-mediated erythroleukemia cell commitment to terminal cell division. *Proc. natn. Acad. Sci. USA*, **80**, 5919–22.

Shen, S. H. & Smithies, O. (1982). Human globin $\psi\beta 2$ is not a globin-related sequence. *Nucleic Acids Res.* **10**, 7809–18.

Shepherd, J. C. W., McGinnis, W., Carrasco, A. E., de Robertis, E. M. & Gehring, W. J., (1984). Fly and frog homoeo domains show homologies with yeast mating type regulatory proteins. *Nature, Lond.* **310**, 70–1.

Shiba, T. & Saigo, K. (1983). Retrovirus-like particles containing RNA homologous to the transposable element *copia* in *Drosophila melanogaster*. *Nature, Lond.* **302**, 119–24.

Shiu-Lee, A., Britten, R. J. & Davidson, E. H. (1977). Short-period repetitive sequence interspersion in cloned fragments of sea urchin DNA. *Cold Spring Harb. Symp. quant. Biol.* **42**, 1065–72.

Shutt, R. H. & Kedes, L. H. (1974). Synthesis of histone mRNA sequences in isolated nuclei of cleavage stage sea urchin embryos. *Cell*, **3**, 283–90.

Simpson, R. T. (1981). Modulation of nucleosome structure by histone subtypes in sea urchin embryos. *Proc. natn. Acad. Sci. USA*, **78**, 6803–7.

Singer, M. F. (1982). SINEs and LINEs: highly repeated short and long interspersed sequences in mammalian genomes. *Cell*, **28**, 433–4.

Sklar, V. E. F., Schwartz, L. B. & Roeder, R. G. (1975). Distinct molecular structures of nuclear class I, II and III DNA-dependent RNA polymerases. *Proc. natn. Acad. Sci. USA*, **72**, 348–52.

Skoultchi, A. & Gross, P. R. (1973). Maternal histone messenger RNA: detection by molecular hybridisation. *Proc. natn. Acad. Sci. USA*, **70**, 2840–4.

Slater, I. & Slater, D. W. (1974). Polyadenylation and transcription following fertilisation. *Proc. natn. Acad. Sci. USA*, **71**, 1103–7.

Slightom, J. L., Blechl, A. E. & Smithies, O. (1980). Human fetal $^{G}\gamma$ and $^{A}\gamma$ globin genes: complete nucleotide sequences. *Cell*, **21**, 627–38.

Smith, M. J., Hough, B. R., Chamberlin, M. E. & Davidson, E. H. (1974). Repetitive and non-repetitive sequences in sea urchin heterogeneous nuclear RNA. *J. molec. Biol.* **85**, 103–26.

Smith, L. & Thorogood, P. (1983). Transfilter studies on the mechanism of epithelio-mesenchymal interaction leading to chondrogenic differentiation of neural crest cells. *J. Embryol. exp. Morph.* **75**, 165–88.

Snyder, M. P., Kimbrell, D., Hunkapiller, M., Hill, R., Fristom, J. & Davidson, N. (1982). A transposable element that splits the promoter region inactivates a *Drosophila* cuticle protein gene. *Proc. natn. Acad. Sci. USA*, **79**, 7430–4.

Sollner-Webb, B., Melchior, Jr, W. & Felsenfeld, G. (1978). DNAase I, DNAase II and staphylococcal nuclease cut at different, yet symmetrically-located, sites in the nucleosome core. *Cell*, **14**, 611–27.

Sollner-Webb, B., Wilkinson, J. A. K., Roan, J. & Reeder, R. H. (1983). Nested control regions promote *Xenopus* ribosomal RNA synthesis by RNA polymerase I. *Cell*, **35**, 199–206.

Sonnenschein, G. E., Geoghegan, T. E. & Brawerman, G. (1976). A major species of mammalian messenger RNA lacking a polyadenylate segment. *Proc. natn. Acad. Sci. USA*, **73**, 3088–92.

Southern, E. M. (1975). Detection of specific sequences among DNA fragments separated by gel electrophoresis. *J. molec. Biol.* **98**, 503–18.

Spadafora, C., Bellard, M., Compton, J. L. & Chambon, P. (1976). The DNA repeat lengths in chromatins from sea urchin sperm and gastrula cells are markedly different. *FEBS Lett.* **69**, 281–5.

Spemann, H. (1928). Die entwicklung seitlicher und dorso-ventraler keimhalfter bei verzogerter kernversorgung. *Z. wiss. Zool.* **132**, 105–34.

Spirin, A. S. (1966). On "masked" forms of messenger RNA in early embryogenesis and in other differentiating systems. In *Current Topics in Developmental Biology*, ed. A. A. Moscona & A. Monroy, vol. 1, pp. 1–38. Academic Press, New York & London.

Spradling, A. C. (1981). The organisation and amplification of two chromosomal domains containing *Drosophila* chorion genes. *Cell*, **27**, 193–201.

Spradling, A. C. & Mahowald, A. P. (1980). Amplification of genes for chorion proteins during oogenesis in *Drosophila melanogaster*. *Proc. natn. Acad. Sci. USA*, **77**, 1096–100.

Spradling, A. C., Penman, S., Campo, M. S. & Bishop, J. O. (1974). Repetitious and unique sequences in the heterogeneous nuclear and cytoplasmic messenger RNA of mammalian and insect cells. *Cell*, **3**, 23–30.

Spritz, R. A., deRiel, J. K., Forget, B. G. & Weissmann, S. M. (1980). Complete nucleotide sequence of the human δ globin gene. *Cell*, **21**, 639–46.

Stalder, J., Groudine, M., Dodgson, J. B., Engel, J. D. & Weintraub, H. (1980a). Hb switching in chickens. *Cell*, **19**, 973–80.

Stalder, J., Larsen, A., Engel, J. D., Dolan, M., Groudine, M. & Weintraub, H.

(1980b). Tissue-specific DNA cleavages in the globin chromatin domain introduced by DNase I. *Cell*, **20**, 451–60.

Stein, J. P., Catterall, J. F., Kristo, P., Means, A. R. & O'Malley, B. W. (1980). Ovomucoid intervening sequences specify functional domains and generate protein polymorphism. *Cell*, **21**, 681–7.

Stein, J. P., Munjaal, R. P., Lagace, L., Lai, E. C., O'Malley, B. W. & Means, A. W. (1983). Tissue-specific expression of a chicken calmodulin pseudogene lacking intervening sequences. *Proc. natn. Acad. Sci. USA*, **80**, 6485–9.

Stein, R., Gruenbaum, Y., Pollack, Y., Razin, A. & Cedar, H. (1982). Clonal inheritance of the pattern of DNA methylation in mouse cells. *Proc. natn. Acad. Sci. USA*, **79**, 61–5.

Stein, R., Sciaky-Gallili, N., Razin, A. & Cedar, H. (1983). Pattern of methylation of two genes coding for housekeeping functions. *Proc. natn. Acad. Sci. USA*, **80**, 2422–6.

Stephenson, E. C., Erba, H. P. & Gall, J. G. (1981). Histone gene clusters of the newt *Notophthalmus* are separated by long tracts of satellite DNA. *Cell*, **24**, 639–47.

Steward, F. C., Mapes, M. O., Kent, A. E. & Holsten, R. D. (1964). Growth and development of cultured plant cells. *Science*, **143**, 20–7.

Stiles, C. D., Lee, K. L. & Kenney, F. T. (1976). Differential degradation of messenger RNAs in mammalian cells. *Proc. nat. Acad. Sci. USA*, **73**, 2634–8.

Struhl, G. (1981a). A homoeotic mutation transforming leg to antenna in *Drosophila*. *Nature, Lond.* **292**, 635–7.

Struhl, G. (1981b). A gene product required for correct initiation of segmental determination in *Drosophila*. *Nature, Lond.* **293**, 36–9.

Struhl, G. (1982). Genes controlling segmental specification in the *Drosophila* thorax. *Proc. natn. Acad. Sci. USA*, **79**, 7380–4.

Struhl, G. (1983). Role of the esc^+ gene product in ensuring the selective expression of segment-specific homoeotic genes in *Drosophila*. *J. Embryol. exp. Morph.* **76**, 297–331.

Struhl, G. (1984a). Splitting the bithorax complex of *Drosophila*. *Nature, Lond.* **308**, 454–6.

Struhl, G. (1984b). A universal genetic key to body plan? *Nature, Lond.* **310**, 10–11.

Struhl, G. & Brower, D. (1982). Early role of the esc^+ gene product in the determination of segments in *Drosophila*. *Cell*, **31**, 285–92.

Sturkie, P. D. & Mueller, W. J. (1976). Reproduction in the female and egg production. In *Avian Physiology*, ed. P. D. Sturkie, 3rd ed., pp. 302–30. Springer-Verlag, New York, Heidelberg, Berlin.

Stryer, L. (1981). *Biochemistry*, 2nd ed. W. H. Freeman & Co., San Francisco.

Sulston, J. E., Albertson, D. G. & Thomson, J. N. (1980). The *Caenorhabditis* male: postembryonic development of non-gonadal structures. *Devl Biol.* **78**, 542–76.

Sulston, J. E., Schierenberg, E., White, J. G. & Thomson, J. N. (1983). The embryonic cell lineage of the nematode *Caenorhabditis elegans*. *Devl Biol.* **100**, 64–119.

Sulston, J. E. & White, J. G. (1980). Regulation and cell autonomy during postembryonic development of *Caenorhabditis elegans*. *Devl Biol.* **78**, 577–97.

Sutcliffe, J. G., Milner, R. J., Bloom, F. E. & Lerner, R. A. (1982). Common 82-nucleotide sequence unique to brain RNA. *Proc. natn. Acad. Sci. USA*, **79**, 4942–6.

Tata, J. R., Hamilton, M. J. & Shields, D. (1972). Effects of amanitin *in vivo* on RNA polymerase and nuclear RNA synthesis. *Nature New Biol.* **238**, 161–4.

Telford, J. L., Kressmann, A., Koski, R. A., Grosschedl, R., Muller, F., Clarkson, S. G. & Birnstiel, M. L. (1979). Delimitation of a promoter for RNA polymerase III by means of a functional test. *Proc. natn. Acad. Sci. USA*, **76**, 2590–4.

Temin, H. M. (1980). Origin of retroviruses from cellular moveable genetic elements. *Cell*, **21**, 599–600.

Teugels, E. & Ghysen, A. (1983). Independence of the numbers of legs and leg ganglia in *Drosophila* bithorax mutants. *Nature, Lond.* **304**, 440–2.

Thireos, G., Griffin-Shea, R. & Kafatos, F. C. (1980). Untranslated mRNA for a chorion protein of *Drosophila melanogaster* accumulates transiently at the onset of specific gene amplification. *Proc. natn. Acad. Sci. USA*, 5789–93.

Thomas, J. O. & Thompson, R. J. (1977). Variation in chromatin structure in two cell types from the same tissue: a short DNA repeat length in cerebral cortex neurons. *Cell*, **10**, 633–40.

Tilghman, S. M., Curtis, P. J., Tiemeier, D. C., Leder, P. & Weissman, C. (1978b). The intervening sequence of a mouse β globin gene is transcribed within the 15S β globin mRNA precursor. *Proc. natn. Acad. Sci. USA*, **75**, 1309–13.

Tilghman, S. M., Tiemeier, D. C., Seidman, J. G., Peterlin, B. M., Sullivan, M., Maizel, J. V. & Leder, P. (1978a). Intervening sequences of DNA identified in the structural portion of a mouse β globin gene. *Proc. natn. Acad. Sci. USA*, **75**, 725–9.

Tobin, A. J., Selvig, S. E. & Lasky, L. (1978). RNA synthesis in avian erythroid cells. *Devl Biol.* **67**, 11–22.

Tolstoshev, P. & Solomon, E. (1982). Collagen genes. *Nature, Lond.* **300**, 581–2.

Treisman, R., Proudfoot, N. J., Shander, M. & Maniatis, T. (1982). A single-base change at a splice site in a β°-thalassaemic gene causes abnormal RNA splicing. *Cell*, **29**, 903–11.

Truman, D. E. S. (1974). *Biochemistry of Cytodifferentiation*. Blackwell Scientific Publications, Oxford.

Truman, D. E. S. (1982). Taxonomies of differentiation. In *Stability and Switching in Cellular Differentiation*, ed. R. M. Clayton & D. E. S. Truman, *Adv. exp. Med. Biol.* vol. 158, pp. 45–53. Plenum Press, New York & London.

Tsai, M. J., Ting, A. C., Nordstrom, J. L., Zimmer, W. & O'Malley, B. W. (1980). Processing of high molecular weight ovalbumin and ovomucoid precursor RNAs to messenger RNA. *Cell*, **22**, 219–30.

Tufaro, F. & Brandhorst, B. P. (1979). Similarity of proteins synthesised by isolated blastomeres of early sea urchin embryos. *Devl Biol.* **72**, 390–7.

Ueda, R. & Okada, M. (1982). Induction of pole cells in sterilised *Drosophila* embryos by injection of subcellular fraction from eggs. *Proc. natn. Acad. Sci. USA*, **79**, 6946–50.

Ursprung, H. (1972). The fine structure of imaginal disks. In *The Biology of Imaginal Disks*, ed. H. Ursprung & R. Nothiger, *Results and Problems in Cell*

Differentiation, vol. 5, pp. 93–107. Springer-Verlag, New York, Heidelberg, Berlin.

Ursprung, H. & Nothiger, R. (eds.) (1972). *The Biology of Imaginal Disks. Results and Problems in Cell Differentiation*, vol. 5. Springer-Verlag, New York, Heidelberg, Berlin.

Van Arsdell, S. W., Denison, R. A., Berstein, L. B., Weiner, A. M., Manser, T. & Gesteland, R. F. (1981). Direct repeats flank three small nuclear RNA pseudogenes in the human genome. *Cell*, **26**, 11–17.

Van den Berg, J., van Ooyen, A., Mantei, N., Schambock, A., Grosveld, G., Flavell, R. A. & Weissman, C. (1978). Comparison of cloned rabbit and mouse β globin genes showing strong evolutionary divergence of two homologous pairs of introns. *Nature, Lond.* **276**, 37–44.

Van Dongen, W. M. A. M., Moorman, A. F. M. & Destrée, O. H. J. (1983). The accumulation of the maternal pool of histone H1A during oogenesis in *Xenopus laevis*. *Cell Differ.* **12**, 257–64.

Vanin, E. F., Goldberg, G. I., Tucker, P. N. & Smithies, O. (1980). A mouse α globin-related pseudogene lacking intervening sequences. *Nature, Lond.* **286**, 222–6.

Varley, J. M., Macgregor, H. C. & Erba, H. P. (1980). Satellite DNA is transcribed on lampbrush chromosomes. *Nature, Lond.* **283**, 686–8.

Verdonk, N. H. (1968). The effect of removing the polar lobe in centrifuged eggs of *Dentalium*. *J. Embryol. exp. Morph.* **19**, 33–42.

Wahli, W. & Dawid, I. B. (1979). Vitellogenin in *Xenopus laevis* is encoded by a small family of genes. *Cell*, **16**, 535–49.

Wahli, W., Dawid, I. B., Ryffel, G. U. & Weber, R. (1981). Vitellogenesis and the vitellogenin gene family, *Science*, **212**, 298–306.

Wahli, W., Dawid, I. B., Wyler, T., Weber, R. & Ryffel, G. U. (1980). Comparative analysis of the structural organisation of two closely related vitellogenin genes in *X. laevis*. *Cell*, **20**, 107–17.

Walker, P. M. B. (1971). Origin of satellite DNA. *Nature, Lond.* **229**, 306–8.

Walker, V. K. & Ashburner, M. (1981). The control of ecdysterone-regulated puffs in *Drosophila* salivary glands. *Cell*, **26**, 269–77.

Waring, G. L., Allis, C. D. & Mahowald, A. P. (1978). Isolation of polar granules and the identification of polar granule specific protein. *Devl Biol.* **66**, 197–206.

Waring, G. L. & Mahowald, A. P. (1979). Identification and time of synthesis of chorion proteins in *Drosophila melanogaster*. *Cell*, **16**, 599–607.

Wasylyk, B., Kedinger, C., Corden, J., Brison, O. & Chambon, P. (1980). Specific *in vitro* initiation of transcription on conalbumin and ovalbumin genes and comparison with adenovirus 2 early and late genes. *Nature, Lond.* **285**, 367–70.

Watson, J. D. (1976). *Molecular Biology of the Gene*, 3rd ed. Benjamin, New York.

Watson, J. D., Tooze, J. & Kurtz, D. T. (1983). *Recombinant DNA: A Short Course*. Scientific American Books, W. H. Freeman & Co, New York.

Weatherall, D. J. & Clegg, J. B. (1979). Recent developments in the molecular genetics of human haemoglobins. *Cell*, **16**, 467–79.

Weigert, M., Perry, R., Kelley, D., Hunkapiller, T., Schilling, J. S. & Hood, L. (1980). The joining of V and J gene segments creates antibody diversity. *Nature, Lond.* **283**, 497–9.

Weil, P. A., Luse, D. S., Segall, J. & Roeder, R. G. (1979). Selective and accurate initiation of transcription at the Ad2 major late promoter in a soluble system dependent on purified RNA polymerase II and DNA. *Cell*, **18**, 469–84.

Weintraub, H. (1978). The nucleosome repeat length increases during erythropoiesis in the chick, *Nucleic Acids Res.* **5**, 1179–88.

Weintraub, H., Larsen, A. & Groudine, M. (1981). α globin switching during the development of chicken embryos: expression and chromatin structure. *Cell*, **24**, 333–44.

Weisbrod, S. (1982). Active chromatin. *Nature, Lond.* **297**, 289–95.

Weisbrod, S. & Weintraub, H. (1979). Isolation of a subclass of nuclear proteins responsible for conferring a DNase I-sensitive structure on globin chromatin. *Proc. natn. Acad. Sci. USA*, **76**, 630–4.

Weisbrod, S. & Weintraub, H. (1981). Isolation of actively transcribed nucleosomes using immobilised HMG 14 and 17 and an analysis of α globin chromatin. *Cell*, **23**, 391–40.

West, M. H. P. & Bonner, W. M. (1982). Histone H2B can be modified by the attachment of ubiquitin. *Nucl. Acids Res.* **8**, 4671–80.

Wieslander, L. & Daneholt, B. (1977). Demonstration of Balbiani ring RNA sequences in polysomes. *J. Cell Biol.* **73**, 260–4.

Welshons, W. V., Lieberman, M. E. & Gorski, J. (1984). Nuclear localization of unoccupied oestrogen receptors. *Nature, Lond.* **307**, 747–9.

Wilcox, M., Brower, D. L. & Smith, R. J. (1981). A position-specific cell surface antigen in the *Drosophila* wing imaginal disc. *Cell*, **25**, 159–64.

Wiley, H. S. & Wallace, R. A. (1981). The structure of vitellogenin. *J. biol. Chem.* **256**, 8626–34.

Wilks, A. F., Cozens, P. J., Mattaj, I. W. & Jost, J. P. (1982). Estrogen induces a demethylation at the 5′ end region of the chicken vitellogenin gene. *Proc. natn. Acad. Sci. USA*, **79**, 4252–5.

Williams, D. L., Wang, S. Y. & Klett, H. (1978). Decrease in functional albumin mRNA during oestrogen-induced vitellogenin biosynthesis in avian liver. *Proc. natn. Acad. Sci. USA*, **75**, 5974–8.

Willing, M. C., Nienhuis, A. W. & Anderson, W. F. (1979). Selective activation of human β but not γ globin gene in human fibroblast × mouse erythroleukemia cell hybrids. *Nature, Lond.* **277**, 534–8.

Wilson, E. B. (1904a). Experimental studies in germinal localisation: I, the germ-regions in the egg of *Dentalium. J. exp. Zool.* **1**, 1–72.

Wilson, E. B. (1904b). Experimental studies in germinal localisation: II, experiments on the cleavage-mosaic in *Patella* and *Dentalium. J. exp. Zool.* **1**, 197–268.

Wiskocil, R., Bensky, P., Dower, W., Goldberger, R. F., Gordon, J. I. & Deeley, R. G. (1980). Coordinate regulation of two estrogen-dependent genes in avian liver. *Proc. natn. Acad. Sci. USA*, **77**, 4474–8.

Wold, B. J., Klein, W. H., Hough-Evans, B. R., Britten, R. J. & Davidson, E. H. (1978). Sea urchin embryo mRNA sequences expressed in the nuclear RNA of adult tissues. *Cell*, **14**, 941–50.

Wolff, E. (1968). Specific interactions between tissues during organogenesis. In *Current Topics in Developmental Biology*, ed. A. A. Moscona & A. Monroy, vol. 3, pp. 65–94. Academic Press, New York & London.

Wolpert, L. (1969). Positional information and the spatial pattern of cellular differentiation. *J. theor. Biol.* **25**, 1–47.

Wood, W. G., Old, J. M., Roberts, A. V. S., Clegg, J. B., Weatherall, D. J. & Quattrin, N. (1978). Human globin gene expression; control of β, δ and $\delta\beta$ chain production. *Cell*, **15**, 437–46.

Woodland, H. R. & Adamson, E. D. (1977). Synthesis and storage of histones during the oogenesis of *Xenopus laevis*. *Devl Biol.* **57**, 118–38.

Woodland, H. R., Flynn, J. M. & Wyllie, A. J. (1979). Utilisation of stored mRNA in *Xenopus* embryos and its replacement by newly synthesised transcripts: histone H1 synthesis using interspecies hybrids. *Cell*, **18**, 165–71.

Wright, S., deBoer, E., Grosveld, F. G. & Flavell, R. A. (1983). Regulated expression of the human β globin gene family in murine erythroleukaemia cells. *Nature, Lond.* **305**, 333–5.

Wu, G. J. (1978). Adenovirus DNA-directed transcription of 5.5S RNA *in vitro*. *Proc. natn. Acad. Sci. USA*, **75**, 2175–9.

Wu, C. & Gilbert, W. (1981). Tissue-specific exposure of chromatin structure of the 5′ terminus of the rat preproinsulin II gene. *Proc. natn. Acad. Sci. USA*, **78**, 1577–80.

Yamada, T. (1977). Control mechanisms in cell-type conversions in newt lens regeneration. In *Monographs in Developmental Biology*, ed. A. Wolsky, vol. 13. S. Karger, Basel.

Young, N. S., Benz, Jr., E. J., Kantor, J. A., Kretschmer, P. & Nienhuis, A. W. (1978). Haemoglobin switching in sheep: only the γ gene is in the active conformation in fetal liver but all the β and γ genes are in the active conformation in bone marrow. *Proc. natn. Acad. Sci. USA*, **75**, 5884–8.

Zanjani, E. D., McGlave, P. B., Bhakthavathsalan, A. & Stamatoyannopoulos, G. (1979). Sheep foetal haemotopoietic cells produce adult haemoglobin when transplanted in the adult animal. *Nature, Lond.* **280**, 495–6.

Zasloff, M. & Felsenfeld, G. (1977). Use of mercury-substituted ribonucleoside triphosphates can lead to artefacts in the analysis of *in vitro* chromatin transcripts. *Biochem. biophys. Res. Commun.* **75**, 598–603.

Zeevi, M., Nevins, J. R. & Darnell, Jr., J. E. (1981). Nuclear RNA is spliced in the absence of poly (A) addition. *Cell*, **26**, 39–46.

Zehner, Z. E. & Paterson, B. M. (1983). Characterisation of the chicken vimentin gene: single copy gene producing multiple mRNAs. *Proc. natn. Acad. Sci. USA*, **80**, 911–15.

Zelenka, P. & Piatigorsky, J. (1976). Reiteration frequency of δ-crystallin DNA in lens and non-lens tissues of chick embryos: δ-crystallin gene is not amplified during lens cell differentiation. *J. biol. Chem.* **25**, 4294–8.

References added in proof

Anderson, K. V. & Nusslein-Volhard, C. (1984). Information for the dorsal-ventral pattern of the *Drosophila* embryo is stored as maternal mRNA. *Nature, Lond.* **311**, 223–7.

Fritton, H. P., Igo-Kemenes, T., Nowock, J., Strech-Jurk, U., Theisen, M. & Sippel, A. E. (1984). Alternative sets of DNase I-hypersensitive sites characterize the various functional states of the chicken lysozyme gene. *Nature, Lond.* **311**, 163–5.

Hafen, E., Kuroiwa, A. & Gehring, W. J. (1984). Spatial distribution of transcripts from the segmentation gene *fushi tarazu* during *Drosophila* embryonic development. *Cell*, **37**, 833–41.

Imaizumi-Scherrer, M. T., Maundrell, K., Civelli, O. & Scherrer, K. (1982). Transcriptional and post-transcriptional regulation in duck erythroblasts. *Devl Biol.* **93**, 126–38.

Lawrence, P. A. & Johnston, P. (1984). The genetic specification of pattern in a *Drosophila* muscle. *Cell*, **36**, 775–82.

North, G. (1984). How to make a fruit fly. *Nature, Lond.* **311**, 214–16.

Renkawitz, R., Schutz, G., von der Ahe, D. & Beato, M. (1984). Sequences in the promoter region of the chicken lysozyme gene required for steroid regulation and receptor binding. *Cell*, **37**, 503–10.

Weissmann, C. (1984). Excision of introns in lariat form. *Nature, Lond.* **311**, 103–4.

Wright, S., Rosenthal, A., Flavell, R. & Grosveld, F. (1984). DNA sequences required for regulated expression of β-globin genes in murine erythroleukemia cells. *Cell*, **38**, 265–73.

Wu, C. (1984). Activating protein factor binds *in vitro* to upstream control sequences in heat shock gene chromatin. *Nature, Lond.* **311**, 81–3.

Index